Combined-Cycle Gas & Steam Turbine Power Plants

3rd Edition

Combined-Cycle Gas & Steam Turbine Power Plants
3rd Edition

Rolf Kehlhofer
Bert Rukes
Frank Hannemann
Franz Stirnimann

Disclaimer: The recommendations, advice, descriptions, and the methods in this book are presented solely for educational purposes. The author and publisher assume no liability whatsoever for any loss or damage that results from the use of any of the material in this book. Use of the material in this book is solely at the risk of the user.

Copyright © 2009 by
PennWell Corporation
1421 South Sheridan Road
Tulsa, Oklahoma 74112-6600 USA

800.752.9764
+1.918.831.9421
sales@pennwell.com
www.pennwellbooks.com
www.pennwell.com

Marketing: Jane Green
National Account Executive: Barbara McGee

Director: Mary McGee
Managing Editor: Stephen Hill
Production Manager: Sheila Brock
Production Editor: Tony Quinn
Cover Designer: Jesse Bennett
Book Designer: Susan Ormston

Library of Congress Cataloging-in-Publication Data
Combined-cycle gas & steam turbine power plants / Rolf Kehlhofer ... [et al.]. –3rd ed.
   p. cm.
Rev. ed. of: Combined-cycle gas & steam turbine power plants / Rolf Kehlhofer.
  Includes bibliographical references and index.
  ISBN 978-1-59370-168-0 (alk. paper)

1. Gas power plants. 2. Cogeneration of electric power and heat. 3. Steam power plants.   I. Kehlhofer, Rolf, 1951- Combined-cycle gas & steam turbine power plants. II. Title: Combined cycle gas and steam turbine power plants.

TK1061.K44 2008
621.31'2132–dc22   2008043560

All rights reserved. No part of this book may be reproduced, stored in a retrieval system, or transcribed in any form or by any means, electronic or mechanical, including photocopying and recording, without the prior written permission of the publisher.

Printed in the United States of America
1 2 3 4 5   13 12 11 10 09

Contents

1. Introduction 1
2. The Electricity Market 5
3. Economics 11
4. Thermodynamic Principles of the Combined-Cycle Plant 35
5. Combined-Cycle Concepts 45
6. Applications of Combined Cycles 135
7. Components 165
8. Control and Automation 211
9. Operating and Part Load Behavior 225
10. Environmental Consideration 261
11. Developmental Trends 277
12. Integrated Gasification Combined Cycle 287
13. Carbon Dioxide Capture and Storage 321
14. Typical Combined-Cycle Plants 349
15. Conclusion 389
Appendix A: Conversions Table 393
Appendix B: Calculation of the Operating Performance of Combined-Cycle Installations 403
Appendix C: Symbols Used 405
Bibliography 407
Index 413
About the Authors 433

List of Figures

Figure 1–1	Simplified flow diagram of a combined cycle	2
Figure 2–1	Market development since 1975 of new power plants sold per year	9
Figure 3–1	Breakdown of the capital requirement for combined-cycle power plant	19
Figure 3–2	The cost of fuels over the years	23
Figure 3–3	Comparison of cost of electricity for base-load operation	29
Figure 3–4	Comparison of cost of electricity for intermediate load	30
Figure 3–5	Influence of fuel cost on the cost of electricity	31
Figure 3–6	Influence of the equivalent utilization time on the cost of electricity	32
Figure 3–7	Influence of interest rate on the cost of electricity	32
Figure 4–1	Temperature/entropy diagrams for various cycles	37
Figure 4–2	Gas turbine inlet temperature (TIT) definitions	38
Figure 4–3	The efficiency of a simple-cycle GT and a combined cycle plant as function of the gas turbine inlet temperature and pressure ratio	42
Figure 4–4a and 4–4b	The efficiency of a simple-cycle GT with single-stage combustion as a function of turbine inlet temperature (TIT) and turbine exhaust temperature	43
Figure 4–5a and 4–5b	The efficiency of a simple-cycle GT with sequential combustion as a function of the turbine inlet temperature (TIT) and the turbine exhaust temperature	44
Figure 5–1	Selection of a combined-cycle power plant concept	46
Figure 5–2	Evolutionary change in combined-cycle design philosophy	47
Figure 5–3	Standardization approach	48
Figure 5–4	Entropy/temperature diagram for a gas turbine process at two different ambient air temperatures	52
Figure 5–5	Relative efficiency of gas turbine, steam process, and combined cycle as function of the air temperature	53
Figure 5–6	Relative power output of a gas turbine, steam turbine, and combined cycle as function of the air temperature	54

Figure 5–7	Relative power output of gas turbine, steam turbine, combined cycle, and relative air pressure versus elevation above sea level	55
Figure 5–8	Relative power output and efficiency of gas turbine and combined cycle as function of relative humidity	56
Figure 5–9	Effect of water and steam injection on relative combined-cycle power	59
Figure 5–10	Effect of condenser pressure on steam turbine output	62
Figure 5–11	Temperature of cooling medium versus condenser pressure for different types of cooling systems	63
Figure 5–12	Flow diagram to show fuel preheating	65
Figure 5–13	Steam turbine output and HRSG efficiency versus gas turbine exhaust temperature for a single-pressure cycle	68
Figure 5–14	Ratio of steam turbine output of a dual pressure compared to a single-pressure cycle as a function of the gas turbine exhaust temperature	69
Figure 5–15	Energy/temperature diagram for an idealized heat exchanger	71
Figure 5–16	Flow diagram of a single-pressure cycle	73
Figure 5–17	Energy/temperature diagram of a single-pressure HRSG	75
Figure 5–18	Heat balance for a single-pressure cycle	76
Figure 5–19	Energy flow diagram for the single-pressure combined-cycle power	77
Figure 5–20	Effect of live steam pressure on steam turbine output for a single-pressure cycle (including steam turbine exhaust moisture content and HRSG efficiency)	78
Figure 5–21	Energy/temperature diagram of a single-pressure HRSG with live-steam pressure of 40 and 105 bar (566 and 1508 psig)	79
Figure 5–22	Effect of live-steam pressure on condenser waste heat at constant condenser pressure	80
Figure 5–23	Effect of live-steam temperature on steam turbine output for a single-pressure cycle with 105 bar (1508 psig) live-steam pressure (including HRSG efficiency and steam turbine exhaust moisture content)	82
Figure 5–24	Effect of pinch point on relative steam turbine power output and relative HRSG heating surface	83

Figure 5–25	Influence of HRSG backpressure on combined-cycle output and efficiency, GT output and efficiency, and HRSG surface	85
Figure 5–26	Effect of feedwater temperature on steam turbine output and HRSG efficiency for cycles with one stage of preheating	87
Figure 5–27	Energy/temperature diagram for a single-pressure HRSG	88
Figure 5–28	Energy/temperature diagram for a conventional boiler	89
Figure 5–29	Flow diagram of a single-pressure cycle with LP preheating loop for high sulfur fuels	92
Figure 5–30	Flow diagram of a dual-pressure cycle for high sulfur fuel	93
Figure 5–31	Effect of feedwater temperature and number of preheating stages on steam turbine output of a dual-pressure cycle	94
Figure 5–32	Flow diagram of a dual-pressure cycle with low sulfur fuel	95
Figure 5–33	Heat balance for a dual-pressure cycle with low sulfur fuel	96
Figure 5–34	Energy flow diagram for a dual-pressure combined-cycle plant	97
Figure 5–35	Energy/temperature diagram for a dual-pressure HRSG	98
Figure 5–36	Effect of the HP and LP pressure on steam turbine output and exhaust moisture content for a dual-pressure cycle	101
Figure 5–37	Effect of LP pressure on HRSG efficiency for a dual-pressure cycle	102
Figure 5–38	Effect of HP and LP steam temperature on steam turbine output for a dual-pressure cycle	103
Figure 5–39	Effect of HP and LP pinch point on steam turbine output and relative HRSG surface for a dual-pressure cycle	104
Figure 5–40	Flow diagram of a triple-pressure cycle	106
Figure 5–41	Heat balance of a triple-pressure cycle	107

Figure 5–42 Energy/temperature diagram
of a triple-pressure HRSG 108

Figure 5–43 Energy flow diagram of a triple-pressure
combined-cycle plant 109

Figure 5–44 Steam turbine output and exhaust moisture content
versus HP and IP pressure for triple-pressure
cycles at constant LP pressure (5 bar) 110

Figure 5–45 Effect of LP pressure on steam turbine output
and relative HRSG surface for triple-pressure
cycles at constant HP (105 bar) and IP (25 bar) 111

Figure 5–46 Live-steam temperature optimization
for a triple-pressure cycle 112

Figure 5–47 Temperature/entropy diagram showing
the effect of "mild reheat" on the steam turbine
expansion line 113

Figure 5–48 Effect of HP and IP evaporator pinch point on steam
turbine output and relative HRSG surface for a
triple-pressure cycle with constant LP pinch point .. 114

Figure 5–49 Flow diagram of a triple-pressure reheat cycle 116

Figure 5–50 Heat balance for a triple-pressure reheat cycle 117

Figure 5–51 Temperature/entropy diagram showing the effect
of full reheat on the steam turbine expansion line .. 118

Figure 5–52 Energy flow diagram for a triple-pressure reheat
combined-cycle plant 118

Figure 5–53 Energy/temperature diagram for a triple-pressure
reheat HRSG 119

Figure 5–54 Steam turbine output and HRSG surface versus HP
and reheat pressure for a triple-pressure reheat cycle
at constant HP and reheat temperature (568/568°C) ... 120

Figure 5–55 Steam turbine output versus HP and reheat steam
temperature for a triple-pressure reheat cycle at constant
HP (120 bar), IP (30 bar) and LP (5 bar) pressure ... 122

Figure 5–56 Flow diagram of a high-pressure reheat cycle
with a HP once-through HRSG and a drum-type
LP section 123

Figure 5–57 Energy/temperature diagram for 647°C (A),
750°C (B) and 1000°C (C) exhaust gas
temperature entering the HRSG 126

Figure 5–58	Effect of temperature after supplementary firing on power output and efficiency relative to that of a single-pressure cycle	127
Figure 5–59	Heat balance for a single-pressure cycle with supplementary firing	128
Figure 5–60	Performance of different combined cycles over the exhaust gas temperature	130
Figure 5–61	Influence of various parameters/measure on combined-cycle output and efficiency	132
Figure 5–62	Impact of HP and reheat steam parameters (pressure and temperature) on combined-cycle net efficiency	133
Figure 6–1	Simplified flow diagram of a cogeneration cycle with a back pressure turbine	137
Figure 6–2	Flow diagram of a cogeneration cycle with an extraction/condensing steam turbine	138
Figure 6–3	Flow diagram of a cogeneration cycle with no steam turbine	139
Figure 6–4	Heat balance for a single-pressure cogeneration cycle with supplementary firing	140
Figure 6–5	Effect of process steam pressure on relative combined-cycle power output and power coefficient for a single-pressure cycle with 750°C supplementary firing	142
Figure 6–6	Effect of power coefficient on electrical efficiency and fuel utilization for a single-pressure cycle with 750°C supplementary firing	143
Figure 6–7	Flow diagram of a cogeneration cycle with a dual-pressure HRSG	144
Figure 6–8	Comparison of 1-stage and 3-stage heating of district heating water	146
Figure 6–9	Heat balance for a cycle with two stages of district heating	147
Figure 6–10	Flow diagram of a district heating/condensing cycle	148
Figure 6–11	Flow diagram of a cycle coupled with a seawater desalination plant	150
Figure 6–12	Flow diagram of a conventional non-reheat steam power plant	152
Figure 6–13	Flow diagram of a combined-cycle plant using an existing steam turbine	152

Figure 6–14	Flow diagram of a gas turbine combined with a conventional steam cycle (fully fired combined-cycle plant)	155
Figure 6–15	Flow diagram of a parallel-fired combined-cycle plant	156
Figure 6–16	Flow diagram of a PFBC process	160
Figure 6–17	Flow diagram of a STIG cycle	161
Figure 6–18	Flow diagram of a turbo STIG cycle	162
Figure 6–19	Flow diagram of a HAT cycle	163
Figure 7–1	Turbine Inlet Temperature	167
Figure 7–2	Industrial Trent derived from the aero Trent 800	169
Figure 7–3	Heavy-duty industrial gas turbine	170
Figure 7–4	Gas turbine with sequential combustion	171
Figure 7–5	Types of losses contributing to overall performance degradation	175
Figure 7–6	Forced circulation heat recovery steam generator	184
Figure 7–7	Natural circulation heat recovery steam generator	185
Figure 7–8	Principle of drum-type and once-through evaporation	187
Figure 7–9	Three-pressure reheat once-through HRSG with drum-type LP and IP sections	188
Figure 7–10	Supplementary fired heat recovery steam generator	193
Figure 7–11	Different single-shaft combined-cycle configurations	197
Figure 7–12	Cross section of a 142 MW reheat steam turbine with a separate HP turbine and a combined IP/LP turbine with axial exhaust	199
Figure 7–13	Cross section of a two-casing steam turbine with geared HP turbine	200
Figure 7–14	Cutaway drawing of an air-cooled generator for use in combined-cycle power plants	202
Figure 7–15	Single-line diagram	203
Figure 7–16	Typical arrangement of an air-cooled condenser	204
Figure 7–17	Typical arrangement of a wet cell cooling tower	205
Figure 7–18	Principle of hybrid cooling tower	206
Figure 8–1	Hierarchic levels of automation	213
Figure 8–2	Standard layout for a modern combined-cycle power plant control room	214
Figure 8–3	Principle diagram for a combined-cycle load control system	216

Figure 8–4	Typical combined-cycle droop characteristic of a GT load controller	218
Figure 8–5	Closed control loops in a combined-cycle plant	219
Figure 9–1	Sliding pressure diagram	227
Figure 9–2	Effect of condenser vacuum on combined-cycle efficiency	231
Figure 9–3	Effect of frequency on relative combined-cycle output and efficiency for full-load operation	232
Figure 9–4	Effect of fuel composition and lower heating value on combined-cycle output and efficiency (base-load, gas-operation with wet cooling tower)	234
Figure 9–5	Part-load efficiency of gas turbine and combined cycle	235
Figure 9–6	Ratio of steam turbine and gas turbine output and live-steam data of a combined-cycle plant at part load	236
Figure 9–7	Part-load efficiency of combined-cycle plant with four single-shaft blocks	238
Figure 9–8	Performance guarantee comparison	241
Figure 9–9	Expected non-recoverable combined-cycle power plant degradation of power output and efficiency with GT operating on clean fuels	242
Figure 9–10	Gas turbine compressor efficiency as function of days operation (or operation hours)	243
Figure 9–11	Relative power increase of a combined-cycle power plant as function of ambient temperature and humidity	247
Figure 9–12	Typical arrangement of a fogging system in the GT air intake	248
Figure 9–13	Air inlet cooling process with chiller in Mollier diagram	250
Figure 9–14	Typical diagram of a chiller system	251
Figure 9–15	Startup curve for a 250–400 MW class combined cycle after eight hours standstill	256
Figure 9–16	Startup curve for a 250–400 MW class combined cycle after 48 hours standstill	257
Figure 9–17	Startup curve for a 250–400 MW class combined cycle after 120 hours standstill	257
Figure 9–18	Combined-cycle shutdown curve	258
Figure 9–19	Switch over diagram from gas to oil operation	259
Figure 10–1	NO_x equilibrium as a function of air temperature	263

Figure 10–2	Flame temperature as a function of the fuel-to-air ratio and combustion air conditions	264
Figure 10–3	NO_x concentration as a function of fuel-to-air ratio and combustion air conditions	265
Figure 10–4	NO_x reduction factor as a function of the water or steam-to-fuel ratio in gas turbines with diffusion combustion	266
Figure 10–5	Cross section of a low NO_x burner	269
Figure 10–6	Cross section of the Siemens dry low NO_x burner	270
Figure 10–7	Heat recovery steam generator with selective catalytic reduction	272
Figure 11–1	Chronology of the gas turbine inlet temperatures based on improved material and cooling technologies	279
Figure 11–2	Geared high-pressure turbine	283
Figure 11–3	Siemens SGT5-8000H gas turbine	284
Figure 11–4	Plant impact for fast cycling capability	285
Figure 12–1	Principle IGCC concept	287
Figure 12–2	Gasification reactions	288
Figure 12–3	Basic technologies of gasification and gasifier technology vendors	291
Figure 12–4	Entrained flow gasifiers	292
Figure 12–5	Schematic of Siemens full-water quench gasifier with water scrubbing	296
Figure 12–6	Schematic of shell coal gasification process with heat recovery and dry fly ash removal	297
Figure 12–7	Regenerable solvent-type AGR process	298
Figure 12–8	Claus process	299
Figure 12–9	Physical absorption process for simultaneous sulfur and CO_2 removal	302
Figure 12–10	ASU process	303
Figure 12–11	Gas turbine syngas conditioning system (Siemens concept)	304
Figure 12–12	Effect of dilution and heating value on NO_x emission	305
Figure 12–13	Effect of dilution and heating value on NO_x emission	306
Figure 12–14	Gas turbine mass flow imbalance	307
Figure 12–15	IGCC concept based on hard coal and Shell gasification and Siemens combined-cycle technology	309
Figure 12–16	Sankey diagram of Shell-based IGCC concept	311

List of Figures • xv

Figure 12–17   IGCC concept based on lignite and HTW gasification and Siemens combined cycle 312
Figure 12–18   IGCC concept with CO_2 capture based on hard coal and Siemens fuel gasifier (SFG) and Siemens combined-cycle technology 314
Figure 12–19   IGCC concept with CO_2 capture based on Shell gasification and Siemens combined-cycle technology 316
Figure 12–20   Sankey diagram of shell-based IGCC concept with CO_2 capture 317
Figure 13–1   Historical trend of energy consumption and global warming 322
Figure 13–2 CO_2-emissions mitigating by fuel switch 323
Figure 13–3   Typical specific CO_2-emissions from fossil fuels 324
Figure 13–4   Typical trend of solubility in physical and chemical washing agents 326
Figure 13–5   Principle of ion transport membrane 328
Figure 13–6   Principle of chemical looping combustion 329
Figure 13–7   Schema of pre-combustion CO_2-capture 329
Figure 13–8   Combined autothermal reforming and CO_2-capture 332
Figure 13–9   Concept of an innovative premix burner for hydrogen-rich syngases 334
Figure 13–10   Schema of post-combustion CO_2-capture 335
Figure 13–11   Post-combustion absorption process 336
Figure 13–12   Post-combustion CO_2 capture under pressure 338
Figure 13–13   Schema of Oxyfuel firing CO_2 capture 339
Figure 13–14   Concept of the Graz cycle 340
Figure 13–15   Direct Oxyfuel fired CO_2 capture, CES cycle 341
Figure 13–16   Principle of the AZEP process 342
Figure 13–17   Principle of the ZESOFC process 343
Figure 13–18   Expected CO_2 avoidance and power generation costs for industrial-scale power plants in operation by 2020 (Source: ZEP) 345
Figure 14–1   View of Taranaki combined-cycle plant 350
Figure 14–2   Process diagram of Taranaki combined-cycle power plant 351
Figure 14–3   Arrangement of Taranaki combined-cycle power plant 353

Figure 14–4	Process diagram for Monterrey combined-cycle power plant	355
Figure 14–5	Layout of the Monterrey combined-cycle power plant	357
Figure 14–6	View of the Phu My 3 combined-cycle power plant	358
Figure 14–7	Process diagram of the Phu My 3 combined-cycle power plant	359
Figure 14–8	Arrangement of the combined-cycle reference power plant SCC5-4000F 2×1	359
Figure 14–9	View of the Palos de la Frontera combined-cycle power plant	361
Figure 14–10	Layout of the Palos de la Frontera combined-cycle power plant	362
Figure 14–11	Process diagram of the Palos de la Frontera combined-cycle power plant (one unit)	363
Figure 14–12	View of the Arcos combined-cycle power plant	365
Figure 14–13	View of Diemen combined-cycle cogeneration plant	368
Figure 14–14	Process diagram of Diemen combined-cycle cogeneration plant	369
Figure 14–15	General arrangement of Diemen combined-cycle cogeneration plant	371
Figure 14–16	View of the Shuweihat S1 plant	372
Figure 14–17	Process diagram of the Shuweihat S1 plant	373
Figure 14–18	Vado Ligure, old layout	375
Figure 14–19	Vado Ligure, new layout	377
Figure 14–20	Thermal process of repowered plant	380
Figure 14–21	CAD Image of the repowered plant	381
Figure 14–22	Flow scheme of Puertollano IGCC plant	382
Figure 14–23	Main technical data of the IGCC Puertollano	384
Figure 14–24	Availability between 1996 and 2003	384
Figure 14–25	Process diagram of the Monthel cogeneration power plant	386
Figure 14–26	Plant arrangement	386
Figure A–1	Calculation of operating and part-load behavior: Method for solving the system of equations	402
	Conversion of the main units used in this book	403
	Conversion formulæ	403

List of Tables

Table 2–1	Installed capacity 2007	7
Table 2–2	World net electricity generation by type 2004 (TWh)	8
Table 2–3	World fuel consumption for fossil-fuelled power plants	8
Table 2–4	Fossil fuels: proved reserves vs. yearly consumption worldwide in 2007	10
Table 3–1	Specific price of various power plants in US$/kW	19
Table 3–2	Net efficiency of various power plants	21
Table 3–3	Fuel flexibility of various power plants	22
Table 3–4	Variable operating and maintenance costs for various power plants of different sizes	24
Table 3–5	Fixed operation and maintenance costs for various power plants of different sizes	25
Table 3–6	Availability and reliability of various power plants	26
Table 3–7	Construction times for various power plants	28
Table 3–8	Inputs for economical comparison	29
Table 4–1	Thermodynamic comparison of gas turbine, steam turbine, and combined-cycle processes	36
Table 4–2	Allowable reduction in steam process efficiency as a function of gas turbine efficiency ($\eta ST = 0.30$)	41
Table 5–1	Comparison of combined-cycle performance data for different cooling systems and ambient temperatures	61
Table 5–2	Possible fuels for combined-cycle applications	66
Table 5–3	Comparison multishaft versus single shaft	67
Table 5–4	Performance comparison of different cycle concepts (natural gas with low sulfur content and GT exhaust gas temperature of 647°C (1197°F)	129
Table 6–1	Conversion of a 500 MW steam turbine power plant into a parallel-fired combined-cycle plant	157
Table 7–1	Main characteristic data of modern gas turbines for power generation	171
Table 7–2	Critical fuel properties	176
Table 7–3	Typical composition of fuel gases for gas turbines	180

Table 7–4	Change of boundary conditions for steam turbines in combined-cycle plants	196
Table 8–1	Operation of steam turbine control loops in a single-shaft combined-cycle plant	222
Table 9–1	Overview of GT air inlet cooling systems	246
Table 9–2	Max ambient temperature and corresponding cooling potentials at different locations	246
Table 9–3	Typical example of combined cycle power improvement with chiller	252
Table 9–4	Specific investment costs for additional output of air inlet cooling system	253
Table 9–5	Expected startup times for a 400 MW combined-cycle plant	255
Table 10–1	Comparison of the heat to be dissipated for various types of 1,000 MW stations	274
Table 12–1	Emission of existing coal-based IGCC	289
Table 12–2	Commercial solid fuel IGCC plants 2007	290
Table 12–3	Allowed chemical impurities of gas turbine fuels (Siemens source)	295
Table 12–4	Investment and efficiency of IGCC concepts	313
Table 13–1	Typical hydrogen-enriched syngases downstream gas cleaning	332
Table 13–2	Characteristics of typical gas turbine fuels	333
Table 13–3	Commercially available CO_2 absorption systems for post-combustion applications	337
Table 13–4	Selection of European CCS demonstration projects (source: The World Energy Book, issue 3)	347
Table 14–1	Combined-Cycle Plant examples—overview	349
Table 14–2	Main technical data of Taranaki combined-cycle power plant	352
Table 14–3	Main technical data of Monterrey combined-cycle power plant	356
Table 14–4	Main technical data of Phu My 3 combined-cycle power plant	360
Table 14–5	Main technical data of Palos de la Frontera combined-cycle power plant	364
Table 14–6	Main technical data of Arcos III combined-cycle power plant	367

Table 14–7 Main technical data of Diemen combined-cycle cogeneration plant 370
Table 14–8 Main technical data of the Shuweihat S1 Independent Water and Power Plant (IWPP) 374
Table 14–9 Main technical data of Vado Ligure combined-cycle power plant 378
Table 14–10 Main technical data of Monthel cogeneration plant 387

1 Introduction

This third edition of the book *Combined-Cycle Gas & Steam Turbine Power Plants* has been updated and extended to give an accurate picture of today's state of this interesting technology for power generation. It also includes chapters on actual themes such as CO_2 capture and storage, as well as integrated gasification combined-cycle plants (IGCC). These topics have gained a lot of attention as part of the discussion around global warming as potential solutions to this issue.

In substance, the book gives a comprehensive overview about the combined-cycle power plant from a thermodynamic, technical, and economical point of view. It is intended to provide material for lectures and provide an excellent understanding of the potential of this technology. Thanks to practical examples, it offers a real help for professional work. It is equally well suited for students interested in power generation.

The book strives to answer to the following two questions:

- What is a combined-cycle power plant?
- Why are combined-cycle plants among the leading technologies for large power plants?

Combined cycle can be defined as a combination of two thermal cycles in one plant. When two cycles are combined, the efficiency that can be achieved is higher than that of one cycle alone. Thermal cycles with the same or with different working media can be combined; however, a combination of cycles with different working media is more interesting because their advantages can complement one another. Normally, when two cycles are combined, the cycle operating at the higher temperature level is called the *topping cycle*. The waste heat it produces is then used in a second process that operates at a lower temperature level, and is therefore called the *bottoming cycle*.

Careful selection of the working media means that an overall process can be created, which makes optimum thermodynamic use of the heat in the upper range of temperatures and returns waste heat to the environment at the lowest temperature level possible. Normally the topping and bottoming cycles are coupled in a heat exchanger.

The combination used today for commercial power generation is that of a gas topping cycle with a water/steam bottoming cycle. Figure 1–1 shows a simplified flow diagram for such a cycle, in which the exhaust heat of a simple cycle gas turbine is used to generate steam that will be expanded in a steam turbine.

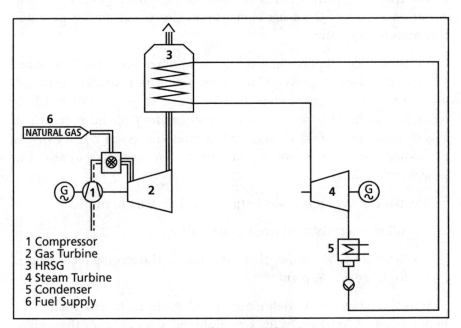

Figure 1–1 Simplified flow diagram of a combined cycle

Replacement of the water/steam in this type of cycle with organic fluids or ammonia has been suggested in the literature because of potential advantages over water in the low exhaust-gas temperature range. As gas turbine exhaust temperatures are increased in line with gas turbine development, however, these advantages become insignificant compared to the high development costs and the potential hazard to the environment through problems such as ammonia leakage. These cycles do not appear likely ever to replace the steam process in a combined-cycle power plant.

The subject of this book is mainly the combination of an open cycle gas turbine with a water/steam cycle. This combination—commonly known as the combined cycle—has several advantages:

- Air is a straightforward medium that can be used in modern gas turbines at high turbine inlet temperature levels (above 1200°C/2192°F), providing the optimum prerequisites for a good topping cycle.

- Steam/water is inexpensive, widely available, non-hazardous, and suitable for medium and low temperature ranges, being ideal for the bottoming cycle.

The initial breakthrough of these cycles onto the commercial power generation market was possible due to the development of the gas turbine. Only since the late 1970s have gas turbine inlet temperatures—and hence, exhaust-gas temperatures—been sufficiently high to allow the design of high-efficiency combined cycles. This breakthrough was made easier because the components of the plant were not new, having already been proven in power plant applications with simple cycle gas turbines and steam turbine processes. This helped to keep development costs low. The result was a power plant with high efficiency, low installation cost, and a fast delivery time.

Today, state of the art combined-cycle plants can reach net efficiency based on the lower heating value (LHV) of the fuel above 58%, and it is only a question of a few years until 60% will be overcome.

2 The Electricity Market

Basic Requirements

The fundamental difference between electricity and most other commodities is that electricity cannot be stored in a practical manner on a large scale. Storing electricity directly is very expensive and can be done only for small quantities (e.g., car batteries). Indirect storage through water or compressed air is more suitable for large-scale applications, but it is very much dependent on topography or geology and is, in many cases, not economical. Actually only storage of water pumped into lakes during off-peak time to be used during peak hours has reached market acceptance.

For this reason, electricity must be produced when the customers need it. It has to be transported by means of extensive transmission and distribution systems, which help to stabilize and equalize the load in the system. Nevertheless, large fluctuations in demand during the day require quick reactions from generation plants to maintain the balance between demand and production.

Fulfillment of this task has been the main focus of the industry from the beginning. A reliable supply of electricity, efficiently delivered, was and is the major priority.

In the last few years a new priority has been set by a global trend to deregulate the electric power market. Deregulation means an opening to competition of what has been a generally closed and protected industry. Private investors have begun to install their own power plants and supply power to the grid. This has created a major focus shift: generators now have to compete to sell their product.

In the past—in a regulated environment—generators sold the power they generated on a cost-plus basis. Production costs were of lower priority than the requirement of grid stability and reliability. Generators focused on reliability through extensive specifications and sufficient system redundancy. Costs could usually be transferred to the end-use customers. The new competitive situation has altered key success factors for the electricity generators.

Today, overall production cost is a key to their success. They must offer electricity at the lowest cost, yet still meet the requirement of flexible adjustment between demand and supply. This cost factor is of major importance for the merchant plants. These are plants built by investors who accept the full market risk, expecting that their assets will be cost competitive, and that they will get a good return on their investment during the lifetime of the plant. Risk and its mitigation carry a much higher weight in this environment.

On one side of this equation, higher competitive risks must be taken to survive in the new markets. On the other side, risks such as cost and schedule overruns have to be minimized. These factors lead us to new behaviors, one of which is that electricity generators are buying plants for fixed, lump-sum prices with guaranteed throughput times. Constructing new plants—with long lead times and high capital costs—are considered to be risky from the private investor's point of view.

Combined-cycle power plants benefit from this change. They have low investment costs and short construction times when compared to large coal-fired stations, and even more so when compared to nuclear plants.

The other benefits of combined cycles are high efficiency and low environmental impact. Worldwide, levels of emissions of all kinds must meet stringent regulations acceptable to the public. It is, therefore, important for power producers to invest in plants with an inherently low level of emissions. Risk mitigation and public acceptance are paramount. Clean plants are easier to permit, build, and operate. Combined-cycle plants—especially those fired with natural gas—are a good choice with their low emissions. (See chapter 10 for more information about this point.)

Throughout this book, we will review different combined-cycle applications, including cogeneration applications; however, more than 80% of the market concerns plants for pure power generation. We will examine how such plants affect and are affected by all of the variables discussed so far in this chapter.

The data cited in table 2–1 are valid for 2007. All figures are rounded to gigawatts (GW). The striking points in this table are:

- Steam turbine plant (STP) for coal-fired application represent about 30% of the total installed capacity worldwide.

- Combined-cycle plants have a total capacity of more than 800 GW, and are approaching a share of 20% compared to less than 5% ten years ago.

Table 2–1 Installed capacity 2007

		GW	
Steam turbines	Coal fired plant	1400	
	Gas fired plant	300	
	Oil fired plant	330	
	Other steam plant	70 [1]	
	Nuclear power plant	395	
	Steam turbine for combined cycle	220	
		Total	2,715
Gas turbines	GT in simple cycle	310	
	GT in combined-cycle	610	
		Total	920
Hydro	Hydro power plant	780	
	Hydro pump storage plant	140	
		Total	920
Diesel		Total	150
New renewables[2]		Total	105
		Total	4,810

Source GS4, BUD 2008-03-07 and IRINA 2008-03-10
1) geothermal, solar, wind, wood and waste   2) coal, gas and oil fired plants

Table 2–2 shows a geographical breakdown of the worldwide electrical energy generated by technology by **terawatt hour** (TWh).

Table 2–2 World net electricity generation by type 2004 (TWh)

	Americas	Asia & Oceania	Europe	Africa	Total
Thermal Plants[1]	3,397.0	5,318.2	1,811.7	407.7	10,934.6
Hydro Plants	1,214.1	917.5	541.7	86.0	2,759.3
Nuclear Plants	902.0	717.7	980.9	14.3	2,615.9
Other[2]	140.1	58.6	140.8	2.0	341.5
Total	5,653.2	7,012.0	3,475.1	510.0	16,650.3

Source: Energy Information Administration: International Energy Annual, October 2007
1) coal, gas and oil fired plants   2) geothermal, solar, wind, wood and waste

The salient points are:

- Two-thirds of the electricity produced worldwide is generated by thermal plants. Coal alone represents more than 40% of the total.
- Combined cycles today generate more than 10%.
- Nuclear and hydro plants are each generating about 15% of the total.
- Other renewables account for about 2%.

Table 2–3 shows the breakdown of the thermal power plants by fuel consumed in 2007.

Table 2–3 World fuel consumption for fossil fuelled power plants

	Worldwide consumption (TTOE[1])	
Oil	283,419	9.6 %
Gas	815,693	27.5 %
Coal	1,862,587	62.9 %
Total	2,961,699	100.0 %

Source: 2005 Energy Balances for World, 2007
1) Thousand Tons of Oil Equivalent

Figure 2–1 shows the evolution of the installed new power plant capacity over the year. Striking are the peaks in 2000 and 2007. The first one is due to a boom in USA where substantial new capacity of combined-cycle plants were installed. The second peak in 2007 is due to large new construction in emerging countries such as China or India as well as in oil producing countries, and also the replacement of old power plants in Europe.

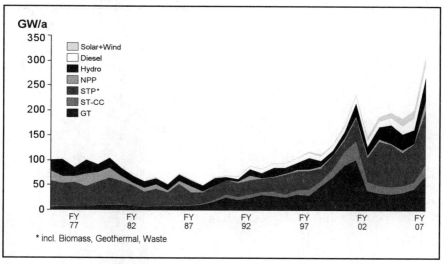

Figure 2–1 Market development since 1975 of new power plants sold per year

The Availability of Fuel

Critical for the future of combined-cycle power plants is the long-term availability of natural gas at reasonable price conditions.

Table 2–4 gives indication about the availability of reserves compared to production today for the main fossil fuels coal, gas and oil.

Taking into account all proven gas reserves, it can be calculated that at the present rate of production the global natural gas reserves will last for approximately 60 years. This estimation shows that natural gas should still be available in the foreseeable future.

Table 2–4 Fossil fuels: proved reserves vs. yearly consumption worldwide in 2007

Fuel	Proved Reserves	Yearly Consumption	Ratio of Proved Reserves to Production
Natural gas	170×10^{12} Nm^3	$3{,}020 \times 10^9$ Nm^3/a	56 a
Oil	$1{,}300 \times 10^9$ barrels	31×10^9 barrels/a	42 a
Coal	910×10^9 ton	6.2×10^9 ton/a	147 a

Source: EIA and BP

3 Economics

In today's mostly liberalized electricity market, the economics of a power plant are evaluated as any other industrial investment.

Typically, an investor sets a target return on an investment; for instance, return on equity (ROE) or internal rate of return (IRR). Similarly to any other investment, this expected return depends on the risk profile of the project. The ROE depends on following factors:

- Market price of the electricity
- Production cost of electricity that includes
  - Interest and amortization cost
  - Fuel cost
  - Operation and maintenance cost

Next to the fuel cost, the **market price of electricity** is usually the biggest unknown factor. The market price in a liberalized economy can fluctuate widely.

Private investors, or independent power producers (IPP), generally try to find electricity off-takers who are buying the power at defined rates under long-term sales contracts (LTSC). Ideally the rate is coupled with the price of fuel, thus removing the most important investment risks.

Thanks to this project structure the investment risk is low, enabling the investor to get access to good credit conditions from financing institutions.

The risks remaining are mostly related to the plant itself: cost, efficiency and reliability.

The extreme case of this type of project is the **Tolling Agreement**. The investor has an agreement with a client, who supplies him with the fuel and takes off the electricity to convert the fuel into electricity at an agreed-upon fee, based on expected efficiency and availability.

The normal power plant project is, however, fully exposed to the market, which leads to a much higher risk profile. In this case, it is important to make a thorough market analysis in order to develop a solid market scenario. This analysis should include sensitivity analyses and worst-case scenarios.

In fossil fuel plants, both electricity and fuel markets have to be analyzed; however, both markets are coupled to a certain extent. For example, if in a given market most of the players are using gas as fuel, the price of electricity will go up if the gas prices are going up and vice versa.

The **interest and amortization cost** has a direct impact on the production cost of electricity produced. The main factors are:

- Debt-to-equity ratio
- Terms of the debt, such as interest and amortization

The debt-to-equity ratio depends mostly on the risk profile of the project. An IPP project with little market risk can be highly leveraged, which means a high percentage of the project financing can be provided in the form of debt. In good projects up to 80% of the capital required can be borrowed, and the equity share is only 20%.

In risky projects, a much lower leverage is possible. Typically, up to 50% of the capital required has to be provided as equity. The normal utility project falls in this category.

Normally, these merchant projects are realized by large utilities with a strong balance sheet. Financing is done based on the balance sheet of the investor (on-balance sheet financing). In this case, the borrower has the balance sheet of the owner of the power plant (the utility) as a guarantee.

On the other side, IPP projects are financed with project financing schemes. Here, a special purpose company (SPC) is created that owns only this power plant. Consequently, the borrower has only the power plants as a guarantee. It is important to have well-structured contracts for construction, operation and maintenance, power sales, and fuel supply to keep the risks at an acceptable level for the lender.

Generally, this type of project is realized by a general contractor under a fixed, lump-sum turnkey contract, also known as an engineering procurement and construction contract (EPC).

Another important factor is the type of power the plant is producing: The value of electricity varies over the year and over the day. Typically, peak power has a much higher value than off-peak power. Consequently, a plant designed to run at base load, such as a nuclear power plant or a coal fired power plant, will in average sell the power at a lower price than a power plant designed to run at peak time.

Typically, one distinguishes among following type of plants.

- Base load (>5000 hrs/a)
- Intermediate load (2000 to 5000 hrs/a)
- Peak load operation (<2000 hrs/a)

Plants with low capital cost but costly fuels are more suitable for peak load operation than those with high capital cost and cheap fuel, which are better suited for base load operation.

Plant cost and flexibility are important factors when defining for which kind of application a given power plant is suitable.

The terms of the financing depend on the market situation and the risk of the selected financing structure. Usually, a project financing structure leads to higher interest rates than a balance sheet financing, provided the borrower has a solid balance sheet.

Production Cost of Electricity

The production costs of electricity are generally expressed in US$/MWh or US cts/kWh, depending on following parameters:

- Capital cost of the project
- Fuel cost
- Operation and maintenance cost

The **capital cost** per unit of electricity for a given power plant depends on following elements:

- Investment costs
- Financing structure (discussed earlier in this chapter)
- Interest rate and return on equity
- Load factor of the plant (or equivalent utilization time)

The **investment costs** are the sum of following positions:

- Power plant contract price(s)
- Interest during construction (depending upon the construction time)
- Owner's cost for the realization of the project (project manager, owner's engineer, land cost, etc.)

The **financing structure** is the defined by the debt-to-equity ratio of the financing and the return on equity is the return expected by the investors on his capital. Both are linked to the risks of the project.

The load factor results from the type of application the plant is intended for: base, intermediate or peak load operation, and the availability and reliability of the power station.

Fuel costs per unit of electricity are proportional to the specific price of the fuel, and inversely proportional to the average electrical efficiency of the installation. This average electrical efficiency must not be mixed up with the electrical efficiency at rated load. It is defined as follows:

$$\bar{\eta} = \eta \cdot \eta_{Oper} \quad (3\text{--}1)$$

where:

η is the electrical net efficiency at rated load. (This is the % of the fuel that is converted into electricity at rated load and new and clean condition)

η_{Oper} is the operating efficiency, which takes into account the following losses:

- start-up and shutdown losses
- higher fuel consumption for part load operation
- aging and fouling of the plant

Operation and maintenance costs consist of fixed costs of operation, maintenance and administration (staff, insurance, etc.), and the variable costs of operation and maintenance (cost of repair, consumables, spare parts, etc.).

The cost of electricity is calculated by adding the capital cost, fuel cost, and operation and maintenance costs.

Present value is generally the basis for economic comparisons among different types of plants. The various costs for a power station are incurred at different times, but for financial calculations are corrected to a single reference time. This is generally the date on which commercial operation starts. These converted amounts are referred to as present value.

The cost of electricity (US$/MWh) is calculated as:

$$Y_{EL} = \frac{TCR \cdot \Psi}{P \cdot T_{eq}} + \frac{Y_F}{\bar{\eta}} + \frac{U_{fix}}{P \cdot T_{eq}} + u_{var} \quad (3\text{--}2)$$

where:

TCR   total capital requirement to be written off (current value of all expenses during planning, procurement, construction, and commissioning, such as the price of the plant, interest during construction, etc.) (US$)

Ψ annuity factor: $\Psi = \dfrac{q-1}{1-q^{-n}}$ [1/a]

P   rated power output in MW

T_{eq} equivalent utilization time at rated power output, in hours per annum (h/a)

Y_F price of fuel (US$/MWh thermal = 3.412 × US$/MBTU)

$\bar{\eta}$ average plant net efficiency

U_{fix} fixed cost of operation, maintenance, and administration

u_{var} variable cost of operation, maintenance, and repair (US$/MWh)

q   1+z

z   average discount rate in percent per annum (%/a)

n   amortization in years

The equivalent utilization time at rated output is the electrical energy generated by a plant in a period of time divided by the electrical energy, which could be produced by the plant running the whole year at rated output (this is a measure for the load factor). This definition enables corrections to be made for the effects of part-load operation, so that different projects can be analyzed on a comparable basis.

For fuel, operation, and maintenance costs, no escalation rates have been applied to calculate the cost of electricity.

For modeling purposes, the cost of electricity can be calculated by using the equivalent utilization time and the fuel price as variables. In a deregulated power generation market, power stations do not sell their electricity based on an average cost of electricity basis, but rather on the basis of demand and supply. Therefore, it is important to understand that the previously referenced equation contains fixed and variable costs in order to avoid producing electricity in times when the sale price is lower than the variable costs.

Fixed costs are:

- Interest and depreciation on capital
- The fixed costs of operation, maintenance and administration (e.g., staff)

Variable costs are:

- The fuel cost
- The variable costs of operation, maintenance and repair (e.g., spare parts)

If the market price of the electricity is falling below the variable cost, the power plant should be shut down because no contribution to the fixed cost can be generated.

Competitive standing of combined-cycle power plants

On the following pages, the combined-cycle power plant is compared with other thermal plants. The comparison evaluates the following types of power stations:

- Steam turbine plants
- Gas turbine plants
- Nuclear plants
- Combined-cycle plants
- Biomass power plants

The range of output ratings under consideration is between 30 and 1250 MW. Combined cycles with a smaller output can, of course, be built, but they are less interesting for pure power generation because the specific costs increase very fast at lower rated output. Applications for heat and power production (e.g., district heating or industrial processes) can make sense.

Comparison of the power plant prices

Table 3–1 shows the specific price of the various types of power plants considered. These prices are valid for a power plant built under a turnkey contract with a fixed, lump-sum price. They are based on prices valid in 2007 assuming progress payment, without interest during construction and other owner's costs.

Table 3–1 Specific price of various power plants in US$/kW

Type of Plant		Specific Price (US$/kW)
Combined Cycle Power Plant	(800 MW)	550–650
Combined Cycle Power Plant	(60 MW)	700–800
Gas Turbine Plant	(250 MW)	300–400
Gas Turbine Plant	(60 MW)	500–600
Steam Power Plant (coal)	(800 MW)	1200–1400
Steam Power Plant (coal)	(60 MW[1])	1000–1200
Nuclear Power Plant	(1250 MW)	2000–3000
Biomass Power Plant	(30 MW)	2000–2500

1) Simple non-reheat plant
Remark: These prices are valid for 2007. They do not include owner's cost and interest during construction.

The data shown indicate trends, so appropriate caution must be taken in applying the data because many factors affect the price of a power plant. These factors include market conditions, site-related factors, types of cooling and related structures, emission limits, labor rates, commercial risks, legal regulations, and so forth.

Table 3–1 also shows the low investment costs required for the gas turbine, which have contributed significantly to its wide-spread acceptance, especially in oil producing countries. Taken together with its simplicity and short start-up time to full load, the gas turbine is an attractive peak-load machine.

Steam power plants are more expensive than combined-cycle power plants. A coal-fired plant, for example, costs approximately two times more than a combined-cycle plant with the same output.

The costs of a nuclear power station are much higher and can vary from one country to the other due to the variety of safety requirements, permitting processes, and other local requirements. In addition, this type of plant has practically disappeared from the European and American markets because of heavy political opposition. A revival seems likely, but it is difficult at this time to give solid information on the cost of nuclear plants.

Figure 3–1 shows the breakdown of the cost of a 400 MW combined-cycle plant according to the different systems. Notice that the gas turbine makes up only one-third of the cost, but delivers two-thirds of the output. If we add the related auxiliary systems and civil work, the cost of the gas turbine island would make up approximately 50% of the total combined-cycle plant cost.

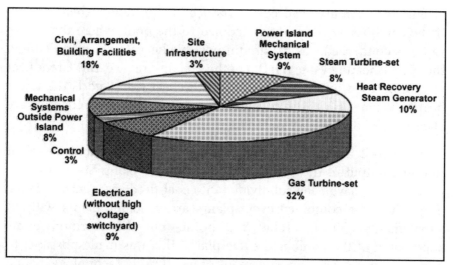

Figure 3–1 Breakdown of the capital requirement for combined-cycle power plant

Comparison of efficiency and fuel costs

At today's fossil fuel prices, efficiency is the most important cost factor for installations operated at intermediate or base load. If an expensive fuel is used, such as natural gas or liquefied natural gas (LNG), the efficiency is crucial. For that reason high efficiency is a prerequisite for an economical fossil-fired plant. The net efficiency of a power station is defined as:

$$\eta = \frac{P_{net}}{\dot{Q}} \qquad (3\text{--}3)$$

where:

P_{net} Power output at the high voltage terminals of the step-up transformer in MW. This number considers the power consumption of all plant auxiliaries

$\dot{Q}$ Heat input to the power station in MJ/s measured at the plant boundary (i.e., mass flow of the fuel multiplied by the lower heating value of the fuel (LHV)

Fuel that contains hydrogen produces water vapor when burned. If the combustion products are cooled to the point where this entire vapor is condensed, the maximum possible heat is extracted defining the higher heating value (HHV) of the fuel. In practice, this latent heat cannot be used in power plants, and the heat extracted is the lower heating value (LHV). The considerations in the following chapters are always based on the LHV of the fuel.

Table 3–2 indicates the electrical efficiency (new and clean) at nominal output of the different types of power plant. Steam turbine power plants have been subdivided into coal-fired and nuclear plants. The values for combined-cycle plants are valid for plants without supplementary firing. Table 3–2 points out the thermodynamic superiority of the combined-cycle plant. This was made possible, to a large extent, by the improvement of gas turbine technology, which already achieves efficiency of 38% to 40% in simple cycle. Only a few years ago, the efficiency of a newly installed coal-fired steam power plant was at these levels, but with much higher investment cost and complexity!

Coal-fired power plants are competitive for base load application in places where cheap coal can be easily transported.

Table 3–2 Net efficiency of various power plants

Type of Plant		Net Efficiency LHV (%)
Combined Cycle Power Plant	(800 MW)	55–59
Combined Cycle Power Plant	(60 MW)	50–54
Gas Turbine Plant	(250 MW)	38–40
Gas Turbine Plant	(60 MW)	35–42
Steam Power Plant (coal)	(800 MW)	(42–) 47
Steam Power Plant (coal)	(60 MW)	30–35
Nuclear Power Plant	(1250 MW)	35
Biomass Power Plant	(30 MW)	28–32

Based on LHV of the fuel at ISO ambient conditions and equipment new and clean

Nuclear power plants have a low efficiency because the light water reactor technology used today allows the generation of steam only with low temperature and pressure. Typically, it is saturated or slightly superheated steam with a pressure of approximately 60 bar (860 psig).

In the future, new technology using, for instance, high temperature reactors will lead to higher efficiency.

Because of its low efficiency, nuclear power needs large cooling towers that are approximately three times as large as for a combined-cycle plant with the same output.

The right selection of the fuel and the corresponding type of power plant is critical for the long-term economical operation of the plant.

The following considerations play a role for the selection of the type of fuel used in a given power plant project:

- Long-term availability of the fuel at a competitive price
- Alternative for the primary fuel as backup
- Risk of supply shortages due to political interference
- Environmental considerations that favor a clean fuel, such as natural gas
- Independence from a single fuel source
- Strategic reasons to use a domestic fuel
- Financing requirements (e.g., uninterruptible fuel supply)

Table 3–3 lists the fuels that can be burned in different types of power plants today.

Table 3–3 Fuel flexibility of various power plants

Fuel	Gas Turbine	Combined Cycle	Steam Power Plant	Biomass Power Plant
Natural gas / LNG	Yes	Yes	Yes[1]	No
Distillate (oil #2)	Yes	Yes	Yes[1]	No
Crude oil / heavy fuel	Yes[2]	Yes[2]	Yes	No
Coal	No	No[4]	Yes	No
Refuse	No	No	Yes	Yes
Biomass	No	No	Yes	Yes
Coal gas, low calorific gas	Yes[3]	Yes[3]	Yes	No
Nuclear fuel	No	No	Yes	No

1) Due to the lower efficiency of steam power plant using this kind of fuel leads to high production costs   2) Heavy oil or crude oil can be burned in some older types of gas turbine   3) These fuels can be used in gas turbines. Modifications to the gas turbine are necessary for fuels with a low heating value, see also chapter 6   4) Can be used in an IGCC plant, see also chapter 12
LNG stands for liquefied natural gas.

Some gas turbines can burn heavy oil or crude oil. Industrial (heavy-duty) gas turbines are more suitable for this type of fuel than those derived from jet technology (aero derivatives). Large combustion chambers are better capable of burning heavy fuels than those with several smaller burners/combustion chambers because the latter are more sensitive to changes in flame length, radiation, and so on. An additional requirement for burning heavy oil or crude oil in a gas turbine is the correct treatment of the fuel, generally by means of cleaning the fuel and/or dosing it with additives. These steps make it possible to remove or inhibit elements that cause high temperature corrosion, such as vanadium and sodium.

It must, however, be mentioned that with today's environmental regulations, the use of heavy (residual) oil in gas turbines is practically impossible in industrialized countries.

Modern gas turbines with high firing temperatures are not designed for heavy fuel operation, but only for natural gas and distillate oil.

As can be seen from tables 3–1 and 3–2, combined-cycle plants are low in capital cost but burn an expensive fuel when compared to coal-fired steam power plants and nuclear plants. The biggest contribution in the cost of the electricity generated is, therefore, the one for the fuel costs. Long-term supply agreements or partnerships with fuel companies, or equity investments by fuel suppliers into the plant, are possible ways to cope with the risk of wide swings in the fuel prices.

Fuel flexibility is higher for a steam turbine power plant than a combined-cycle power plant. But combined-cycle plants are superior to steam power plants for power generation when gas or diesel oil is used, due to higher efficiency and lower investment costs.

As seen in figure 3–2, the prices of fossil fuels are highly volatile, especially gas and oil. This volatility must be duly considered during the economic evaluation of a project and, as already mentioned, it is important to understand how far the market prices of electricity are coupled with the prices of the fuel selected.

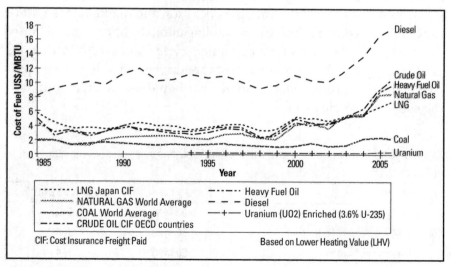

Figure 3–2 The cost of fuels over the years

Uranium, the fuel for nuclear power stations, and coal, the main fuel for steam power plants, have comparatively low costs. Combined-cycle power plants mainly use natural gas that is more expensive and generally more volatile than coal. Since 2004 all the prices of all fuels have increased substantially. Despite this apparent disadvantage, combined-cycle power plants are very popular with investors because they:

- Efficiently convert fuel into electricity
- Have a low impact on the environment
- Have high operation flexibility (startup time, high-load gradient)
- Have a short construction time
- Are are easier to permit than other types of plants

Comparison of operation and maintenance costs

At current levels of fuel and capital cost, operation and maintenance costs affect the economy of a power plant only in a limited manner. They account for less than a tenth of the cost of electricity in a combined-cycle plant. Table 3–4 shows the variable operation and maintenance cost of the different power plants (cost depending upon the operation regime of the plant). Variable costs for a combined-cycle plant are lower than for gas turbine plants because these costs are driven by the high spare part cost of the gas turbine, which can be distributed over a larger output in the combined-cycle plant.

Table 3–4 Variable operating and maintenance costs for various power plants of different sizes

Type of Plant		Cost in US$/MWh
Combined Cycle Power Plant	(800 MW)	2–3
Combined Cycle Power Plant	(60 MW)	3–4
Gas Turbine Power Plant	(250 MW)	3–4
Gas Turbine Power Plant	(60 MW)	4–5
Steam Power Plant (coal)	(800 MW)	2.5–3.5
Nuclear Power Plant	(1250 MW)	2.0
Biomass Power Plant	(30 MW)	5–8

Table 3–5 shows the fixed operation and maintenance costs of the different power plants (cost independent of the operating regime). These are mostly personnel and insurance costs.

Table 3–5 Fixed operation and maintenance costs for various power plants of different sizes

Type of Plant		Cost in Mio. US$/a
Combined Cycle Power Plant	(800 MW)	6–8
Combined Cycle Power Plant	(60 MW)	3–4
Gas Turbine Power Plant	(250 MW)	2–2.5
Gas Turbine Power Plant	(60 MW)	1–1.5
Steam Power Plant (coal)	(800 MW)	12–15
Nuclear Power Plant	(1250 MW)	40–60
Biomass Power Plant	(30 MW)	3–4

Remark: These costs include the insurance cost.

Nuclear power stations and conventional steam turbine power plants require a much larger staff than for other kinds of plants, which explains the high fixed costs of these types of power plant.

Comparison of availability and reliability

The availability of a power plant is defined as:

$$AF = \frac{PH - SOH - FOH}{PH} \qquad (3\text{–}4)$$

where:

- PH   hours of the period considered (normally one year) which amounts to 8,760h
- SOH scheduled outage hours for planned maintenance
- FOH forced outage hours for unplanned outages and repairs

The reliability of a power plant is defined as:

$$RF = \frac{PH - FOH}{PH} \qquad (3\text{–}5)$$

So reliability is the percentage of the time between planned overhauls where the plant is generating or is ready to generate electricity, whereas the availability is the percentage of total time where power could be produced.

Availability and reliability have a big impact on plant economy. When a unit is down, power must be generated in another power station or purchased from another producer. In each case, replacement power is generally more expensive. The power station's fixed costs are incurred whether the plant is running or not.

In deregulated markets, reliability is crucial. At peak tariff hours, a major portion of the income is generated and the plant must be reliable. Scheduled outages can be planned for off-peak periods when tariffs are close to or even below variable costs. Then only a small income loss results from the planned outages.

Typical average figures for the availability and reliability of well designed and maintained plants are indicated in table 3–6.

Table 3–6 Availability and reliability of various power plants

Type of Plant	Availability	Reliability
Combined cycle plant	90–94%	95–98%
Gas turbine plant (gas fired)	90–95%	97–99%
Steam turbine plant (coal fired)	88–92%	94–98%
Nuclear power plant	88–92%	94–98%

These figures are valid for plants operated at base load. They would be lower for peak- or intermediate-load operation because frequent startups and shutdowns reduce the lifetime of critical components and increase the scheduled maintenance and forced outage rates.

The major factors determining plant availability and reliability are:

- Design of the major components
- Engineering of the plant as whole, especially of the interfaces between the systems

- Mode of operation (whether base-, intermediate-, or peak-load duty)
- Type of fuel
- Qualifications and skill of the operating and maintenance staff
- Adherence to manufacturer's operating and maintenance instructions (preventive maintenance)

A high availability has a positive impact on the cost of electricity because it allows an operator to run a power plant with a higher utilization time per year and, therefore, achieve a higher income.

Comparison of construction time

The time required for construction affects the economics of a project. The longer it takes, the larger the capital employed because interest, insurance, and taxes during the construction period add to the price of the plant. It also creates an additional risk for the investor because the market forecast has to look more in the future than for a plant with a shorter realization time.

Table 3–7 shows the time required to build the various types of power plants.

A gas turbine in a simple-cycle application can be installed within the shortest time frame because of its standardized design. Gas turbines, therefore, help secure power generation in fast-growing economies. Additional time is needed for the completion of a combined-cycle plant. Combined-cycle plants can be installed in two phases, with the gas turbine running first in simple-cycle mode, and then in combined-cycle mode as the steam cycle becomes available. With this procedure, two-thirds of the power is available in the time required for a gas turbine installation. However, an outage of two to three months is needed to convert the gas turbine power plant from simple cycle to combined cycle.

Table 3–7 Construction times for various power plants

Type of Plant	Time in Months[3]
Combined cycle power plant[2]	20–30
Gas turbine plant[2]	12–24
Steam power plant (coal)	40–50
Nuclear power plant	60–80[1]
Biomass power plant	22–26

1) Very difficult to give good figures due to the lack of actual experience at least in Europe and USA 2) Depends very much on the market situation 3) From notice to proceed to commercial operation

Besides the actual construction time of a power plant, the time elapsing between start of the project until financial closing and its predictability are also important factors to be considered in the development of a project.

The time for the development of a project varies widely from one project to the other due to local factors. However, it is a fact that gas-fired combined-cycle power plants are easier to develop than coal-fired plants. The main reason is the easier permitting and lower risk of opposition against this kind of project.

Comparison of the cost of electricity

Based on the data presented in table 3–8 and equation (3–2), the cost of electricity has been calculated for various parameters of the following plants.

- Combined-cycle power plant (800 MW)
- Gas turbine plant (250 MW)
- Steam-power plant (coal) (800 MW)
- Nuclear-power plant (1250 MW)

Table 3–8 Inputs for economical comparison

Type of Plant	Description		Investment cost[1] in US$/kW	Average efficiency (LHV) in %	Fuel price[2] in US$/MBTu	Depreciation[1] time in [a]
Combined cycle power plant	(800 MW)	2 × GT 1 × ST	750	56.5	8.0	25
Gas turbine plant	(250 MW)	1 × GT	413	37.5	8.0	25
Steam power plant coal	(800 MW)	1 × ST	1716	44	3.5	25
Nuclear power plant	(1250 MW)	1 × ST	3500	34.5	0.5	40

1) Including interest during construction and owner's cost   2) Fuel price based on LHV

Figure 3–3 shows the cost of electricity for the various plants running in base load (8000 hrs/a). The costs are broken down in fuel, operation and maintenance (O&M), and capital cost. The gas-fired plants, combined-cycle plants, and gas turbine plants show high fuel cost and comparatively low capital cost. On the other side, we have the nuclear power plant with its high capital cost but very low fuel cost. In between, we have the coal-fired plant.

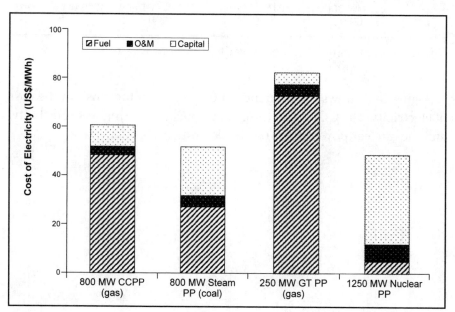

Figure 3–3 Comparison of cost of electricity for base load operation

Figure 3–4 shows the same comparison but for an equivalent utilization time of only 4000 hrs/a. Here the combined-cycle power plants (CCPPs) show the lowest cost of electricity and the nuclear power plant the highest. Out of these figures, the obvious conclusion is that CCPPs are best for intermediate load, and coal-fired or nuclear power plants are well suited for base load application.

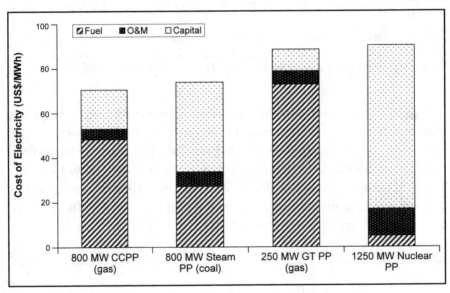

Figure 3–4 Comparison of cost of electricity for intermediate load

Figure 3–5 shows the influence of the charge of fuel cost on the cost of electricity. The CCPP and the gas turbine have the highest sensitivity, and the nuclear power plant has the lowest.

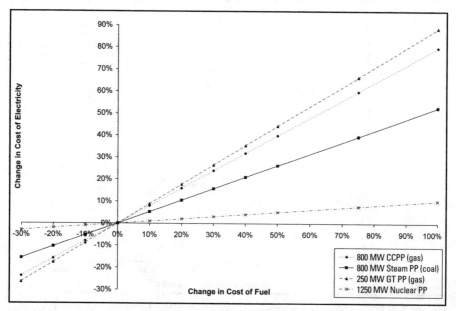

Figure 3–5 Influence of fuel cost on the cost of electricity

The influence of the equivalent utilization time on the cost of electricity is shown in figure 3–6. This figure shows the following best technology:

- Below 1500 hrs/a   gas turbine
- Between 1500 and 5000 hrs/a   CCPPs
- Between 5000 and 8000 hrs/a   coal-fired plant
- Above 7000 hrs/a   nuclear power plant

Figure 3–7 shows the influence of the interest rate on the cost of electricity. In this case, the nuclear power plant due to its large investment cost shows the highest sensitivity. On the other side, with a low sensitivity, we have the CCPP and gas turbine.

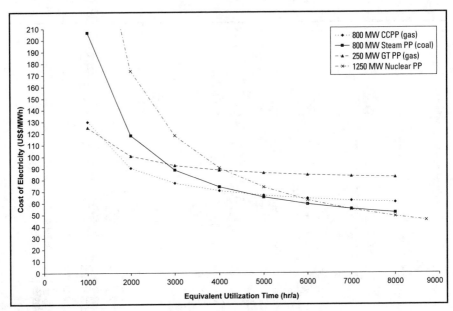

Figure 3–6 Influence of the equivalent utilization time on the cost of electricity

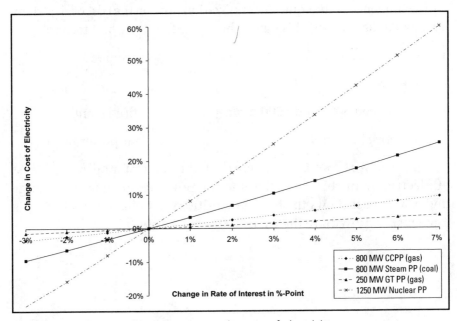

Figure 3–7 Influence of interest rate on the cost of electricity

The CO_2 emissions of the plant are having a more direct impact on the economics of a power plant due to the effort to globally limit these kinds of emissions. If we compare the various plants under this aspect, we get the following picture:

The combined-cycle plant emits about 40% of the CO_2 of a coal-fired plant.

Nuclear power plants or biomass plants are obviously still better because they produce nearly no additional CO_2 (see chapter 13).

No cost for CO_2 emissions were considered in the previous calculations.

Efficiency improvements

As noted, combined-cycle power plants are generally fired with natural gas, which is more expensive than uranium or coal. Consequently, the efficiency of the conversion of fuel into electricity is very important. The question of how much can additionally be invested in a combined-cycle plant to gain additional electrical efficiency is answered as follows.

The maximum additional capital that can be invested to increase the efficiency by 1% is given by the limit at which the cost of electricity remains constant.

Using equation (3–2) to determine the cost of electricity, the following calculations can be done.

$$Y_{EL} = \frac{TCR_1 \cdot \Psi}{P_1 \cdot T_{eq}} + \frac{Y_F}{\eta_1} + \frac{U_{FIX1}}{P_1 \cdot T_{eq}} + u_{var1}$$

$$= \frac{TCR_2 \cdot \Psi}{1.01 \cdot P_1 \cdot T_{eq}} + \frac{Y_F}{1.01 \cdot \overline{\eta}_1} + \frac{U_{FIX2}}{1.01 \cdot P_1 \cdot T_{eq}} + u_{var2} \quad (3-6)$$

Index 1   is used for the plant that is used as the reference

Index 2   is used for the plant with a 1% improved efficiency and 1% improved output

Because operation and maintenance costs are approximately only a tenth of the cost of electricity and remain for both plants, the previous equation can be simplified to:

$$\frac{TCR_1 \cdot \Psi}{P_1 \cdot T_{eq}} + \frac{Y_F}{\overline{\eta}_1} \approx \frac{TCR_2 \cdot \Psi}{1.01 \cdot P_1 \cdot T_{eq}} + \frac{Y_F}{1.01 \cdot \overline{\eta}_1} \qquad (3\text{--}7)$$

which can be solved for TCR_2.

$$TCR_2 \approx TCR_1 \cdot 1.01 + \left(\frac{Y_F}{\overline{\eta}_1} - \frac{Y_F}{1.01 \cdot \overline{\eta}_1}\right) \frac{1.01 \cdot P_1 \cdot T_{eq}}{\Psi} \qquad (3\text{--}8)$$

with:

TCR_1	=	600 million U.S. $
Y_F	=	8.0 U.S. $ /MBTU
$\overline{\eta}_1$	=	56.5%
Ψ	=	9.37 (25 years/8%)
P_1	=	800 MW
T_{eq}	=	5000 hrs/a
TCR_2	=	626.6 million US$ = 1.044 × TCR_1

In this example, for a 1% efficiency increase, 4.4% more capital can be invested. It can be seen from previous equation that this percentage will increase if the fuel price is higher or a lower discount rate is used. A smaller percentage results for fewer operating hours per year or a shorter depreciation period.

These considerations are essential in optimizing a combined-cycle plant for a given application.

4 Thermodynamic Principles of the Combined-Cycle Plant

Basic Considerations

The Carnot efficiency is the maximum efficiency of an ideal thermal process:

$$\eta_c = \frac{T_E - T_A}{T_E} \qquad (4\text{--}1)$$

where:

- η_c = Carnot efficiency [%]
- T_E = Temperature of the energy supplied [K]
- T_A = Temperature of the environment [K]

Naturally, the efficiencies of real processes are lower because there are losses involved. A distinction is drawn between energetic and exergetic losses. Energetic losses are mainly heat losses, and are thus energy that is lost from the process. Exergetic losses are internal losses caused by irreversible processes in accordance with the second law of thermodynamics.

The process efficiency can be improved by raising the maximum temperature in the cycle, releasing the waste heat at a lower temperature, or improving the process to minimize the internal exergetic losses.

The interest in combined cycles arises particularly from these considerations. By its nature, no single cycle can make both improvements to an equal extent. It thus seems reasonable to combine two cycles—one with high process temperatures, and the other with a good *cold end*.

In a simple-cycle gas turbine, attainable process temperatures are high as energy is supplied directly to the cycle without heat exchange. The exhaust heat temperature, however, is also quite high. In the steam cycle, the maximum process temperature is much lower than in the gas turbine process, but the exhaust heat is returned to the environment at a low temperature. As illustrated in table 4–1, combining a gas turbine and a steam turbine thus offers the best possible basis for a high efficiency thermal process.

The last line in the table shows the Carnot efficiencies of the various processes (i.e., the efficiencies that would be attainable if the processes took place without internal exergetic losses). Although that is not the case in reality, this figure can be used as an indicator of the quality of a thermal process. The value shown makes clear just how interesting the combined-cycle power plant is when compared to processes with only one cycle. Even a sophisticated supercritical conventional reheat steam turbine power plant has a Carnot efficiency around 20% lower than that of a good combined-cycle plant.

For combined-cycle power plants actual plant efficiencies are around 75% of the Carnot efficiency, whereas for conventional steam power plants this figure is around 80%. The differences between the actual efficiencies attained with a combined-cycle power plant and the other processes are, therefore, not quite as large as illustrated in table 4–1. The relatively larger drop in the combined-cycle efficiency is caused by higher internal energy losses due to the temperature differential for exchanging heat between the gas turbine exhaust and the water/steam cycle.

Table 4–1 Thermodynamic comparison of gas turbine, steam turbine and combined-cycle processes

	GT	ST	CC
Average temperature of heat supplied, K (°C)	1000–1350 (727–1078)	640–700 (368–428)	1000–1350 (727–1078)
Average temperature of exhaust heat, K (°C)	550–600 (278–328)	300–350 (28–78)	300–350 (28–78)
Carnot efficiency, %	45–55	45–57	65–78

GT = Gas Turbine Power Plant, ST = Steam Turbine Power Plant, CC = Combined Cycle Power Plant

As shown in figure 4–1, which compares the temperature/entropy diagrams of the processes, the combined cycle best utilizes the temperature differential in the heat supplied despite an additional exergetic loss between the gas and the steam process.

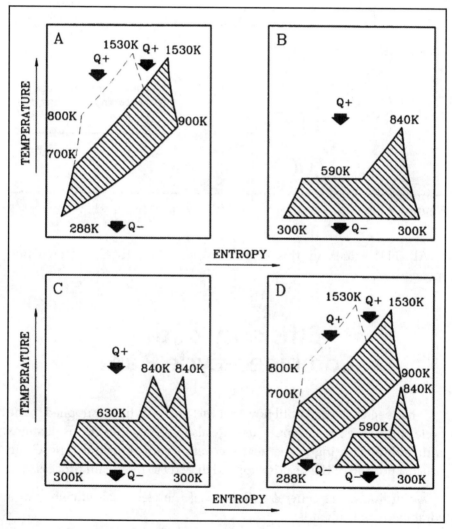

Figure 4–1 Temperature/entropy diagrams for various cycles

Definitions of gas turbine inlet temperature (TIT) are shown in figure 4–2.

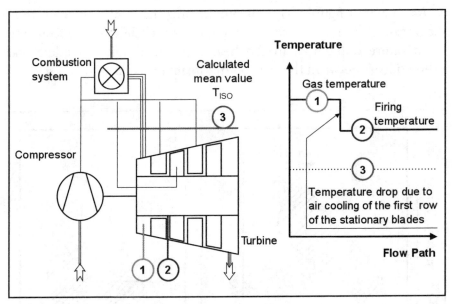

Figure 4–2 Gas turbine inlet temperature (TIT) definitions

All TIT values in this book are according to T_{ISO} (calculated mean value).

Efficiency of the Combined-Cycle Plant

It has been assumed until now that fuel energy is being supplied to the cycle only in the gas turbine. There are also combined-cycle installations with additional firing in the heat recovery steam generator (HRSG), in which a portion of the heat is supplied directly to the steam process.

Accordingly, the general definition of the electrical efficiency of a combined-cycle plant is:

$$\eta_{CC} = \frac{P_{GT} + P_{ST}}{\dot{Q}_{GT} + \dot{Q}_{SF}} \tag{4-2}$$

where:

P_{GT} Gas turbine output

P_{ST} Steam turbine output

$\dot{Q}_{GT}$ Gas turbine fuel consumption

$\dot{Q}_{SF}$ Additional/supplementary firing fuel consumption

This equation shows the gross efficiency of the combined cycle because no station service power consumption and electrical losses, also called auxiliary consumption (P_{Aux}), have been deducted. If station auxiliary consumption is considered the net efficiency of the combined cycle is given by:

$$\eta_{CC,net} = \frac{P_{GT} + P_{ST} - P_{Aux}}{\dot{Q}_{GT} + \dot{Q}_{SF}} \qquad (4\text{--}3)$$

In general, the efficiencies of the simple cycle gas and steam turbine processes can be defined in a similar manner:

$$\eta_{GT} = \frac{P_{GT}}{\dot{Q}_{GT}} \qquad (4\text{--}4)$$

$$\eta_{ST} = \frac{P_{ST}}{\dot{Q}_{GT,Exh} + \dot{Q}_{SF}} \qquad (4\text{--}5)$$

where:

$$\dot{Q}_{GT,Exh} \cong \dot{Q}_{GT}(1 - \eta_{GT}) \qquad (4\text{--}6)$$

Combining these two equations yields:

$$\eta_{ST} = \frac{P_{ST}}{\dot{Q}_{GT}(1 - \eta_{GT}) + \dot{Q}_{SF}} \qquad (4\text{--}7)$$

This equation expresses the steam process efficiency of the combined cycle.

If there is no supplementary firing in the HRSG, equations (4–2) to (4–7) can be simplified by eliminating $\dot{Q}_{SF}$, ($\dot{Q}_{SF} = 0$). In view of earlier considerations, it is generally better to burn the fuel directly in a modern gas turbine rather than in the HRSG because the temperature level at which heat is supplied to the process is higher (GT versus ST process in table 4–1). For that reason, the utilization of supplementary firing is decreasing.

The factors involved in combined-cycle installations with supplementary firing are discussed in more detailed in chapter 5.

Efficiency of combined cycles without supplementary firing in the HRSG

The most common and straightforward type of combined cycle is one in which fuel is supplied in the gas turbine combustion chamber without additional heat supplied in the HRSG. By substituting equations (4–4) and (4–7) into equation (4–2):

$$\eta_{CC} = \frac{\eta_{GT} \cdot \dot{Q}_{GT} + \eta_{ST} \cdot \dot{Q}_{GT}(1 - \eta_{GT})}{\dot{Q}_{GT}}$$

$$= \eta_{GT} + \eta_{ST}(1 - \eta_{GT}) \tag{4–8}$$

Differentiation makes it possible to estimate the effect that a change in efficiency of the gas turbine has on overall efficiency:

$$\frac{\partial \eta_{CC}}{\partial \eta_{GT}} = 1 + \frac{\partial \eta_{ST}}{\partial \eta_{GT}}(1 - \eta_{GT}) - \eta_{ST} \tag{4–9}$$

Increasing the gas turbine efficiency improves the overall efficiency only if:

$$\frac{\partial \eta_{CC}}{\partial \eta_{GT}} > 0 \tag{4–10}$$

From equation (4–9):

$$-\frac{\partial \eta_{ST}}{\partial \eta_{GT}} < \frac{1 - \eta_{ST}}{1 - \eta_{GT}} \tag{4–11}$$

> **Improving the gas turbine efficiency is helpful only if it does not cause too great a drop in the efficiency of the steam process.**

Table 4–2 indicates that when the efficiency of the gas turbine is raised, the reduction in efficiency of the steam process may be greater. For example, steam process efficiency can be reduced from 30% to 27.8% (30/1.08) in case the gas turbine efficiency is raised from 30% to 35% to keep the overall combined-cycle efficiency.

Table 4–2 Allowable Reduction in Steam Process Efficiency as a Function of Gas Turbine Efficiency (η_{ST} = 0.30)

η_{GT}	0.3	0.35	0.4
$-\dfrac{\partial \eta_{ST}}{\partial \eta_{ST}}$	1.0	1.08	1.17

The proportion of the overall output being provided by the gas turbine increases, reducing the effect of lower efficiency in the steam cycle. But a gas turbine with maximum efficiency still does not always provide an optimum combined-cycle plant. For example, for a gas turbine with single stage combustion at constant turbine inlet temperature, a very high-pressure ratio attains a higher efficiency than a moderate pressure ratio. However, the efficiency of the combined-cycle plant with the second machine is normally better because the steam turbine operates far more efficiently with the higher exhaust gas temperature and produces a greater output.

Figure 4–3 shows that in a simple-cycle (GT) power plant, the efficiency is driven by the pressure ratio. However, in case of a combined-cycle plant, the efficiency is driven by the gas turbine inlet temperature (TIT).

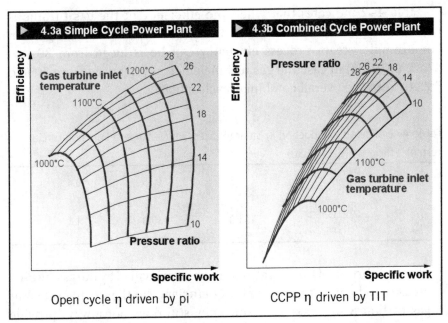

Figure 4–3 The efficiency of a simple-cycle GT and a combined-cycle plant as function of the gas turbine inlet temperature and pressure ratio

Figure 4–4a demonstrates the efficiency of the simple-cycle gas turbine with single-stage combustion as a function of the turbine inlet and exhaust gas temperatures. The maximum efficiency is reached when the exhaust gas temperatures are quite low. In this case a low exhaust temperature is equivalent to a high-pressure ratio.

Figure 4–4b shows the overall efficiency of the combined cycle based on the same gas turbine. Compared to figure 4–4a, the optimum point has shifted toward higher exhaust temperatures from the gas turbine, which again indicates an over-proportional improvement of the water/steam cycle when compared to the loss in gas turbine efficiency. For economic reasons, current gas turbines are generally optimized with respect to maximum power density (output per unit air flow) rather than efficiency. Often, this optimum coincides fairly accurately with the optimum efficiency of the combined-cycle plant. As a result, most of today's gas turbines are optimally suited for combined-cycle installations.

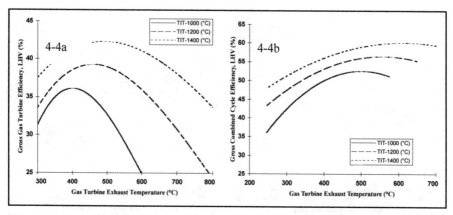

Figure 4–4a and **4–4b** Efficiency of a simple-cycle GT with single stage combustion and combined cycle as a function of turbine inlet temperature (TIT) and turbine exhaust temperature

Gas turbines of a more complicated design (i.e., with intermediate cooling in the compressor or recuperator) are less suitable for combined-cycle plants. They normally have a high simple cycle efficiency combined with a low exhaust gas temperature, so that the efficiency of the water/steam cycle is accordingly lower.

The effects of high-pressure ratio and low combined-cycle efficiency can be decoupled if the gas turbine is designed with sequential combustion (air, upon leaving the compressor is passed through the first combustion chamber and expands in the first turbine stage before final combustion and expansion). Gas turbines with sequential combustion have practically the same simple-cycle efficiencies as single-combustion gas turbines at the same overall pressure ratio.

For comparison the same curves shown in figure 4–4a and 4–4b are shown for gas turbines with sequential combustion in figure 4–5a and 4–5b. Figure 4–5a shows almost the same simple cycle gas turbine efficiency level for the same gas turbine inlet temperatures (TIT). However, the exhaust gas temperatures are substantially higher, clearly improving the combined-cycle efficiency levels of figure 4–5b when compared to figure 4–4b. As for figure 4–4a, a low exhaust gas temperature is equivalent to a high pressure ratio of the sequential part (low pressure part) of the gas turbine. For consistency, the pressure ratio of the high-pressure turbine is kept constant at a ratio of 1.7:1.

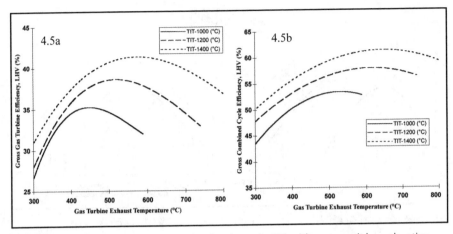

Figure 4–5a and **4–5b** Efficiency of a simple-cycle GT with sequential combustion and combined cycle as a function of turbine inlet temperature (TIT) and turbine exhaust temperature

The main advantage of a sequential-fired gas turbine is that the draw back of single-combustion gas turbines (i.e., pressure ratio and exhaust gas temperature) is eliminated through the reheat process. This gives the ideal base for improved combined-cycle efficiencies, which also fits well to the Carnot comparison.

In summary, it may be said that the optimum gas turbine for simple cycle and the optimum gas turbine for combined-cycle plants are not the same. The gas turbine with the highest efficiency does not necessarily produce the best overall efficiency of the combined-cycle plant.

The type of gas turbine (i.e., gas turbine concept) and turbine inlet temperature are important factors. Similar considerations also apply with regard to the efficiency of the steam cycle. The gas turbine is generally a standard machine, and must, therefore, be optimized by the manufactures for its main application (i.e., combined cycle or simple cycle). The exhaust heat available for the steam process is thus given, and the challenge lies mainly in a cost effective conversion into a mechanical energy.

5 Combined-Cycle Concepts

Selection of the Combined-Cycle Power Plant Concept

This section shows what factors and process steps must be considered when choosing a combined-cycle power plant concept (see figure 5–1). There are three main steps in this process:

1. Analyze the customer requirements

2. Assess the site related factors/influence of the site conditions

3. Determine the plant concept solution

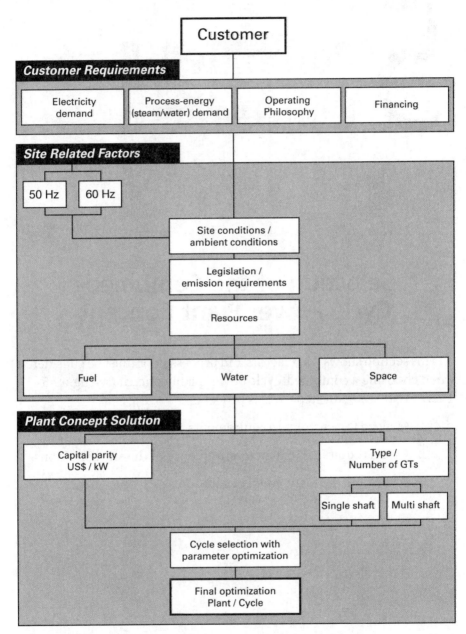

Figure 5–1 Selection of a combined-cycle power plant concept

For large combined-cycle power plants (>200 MW) with the main goal being to produce electricity in an efficient manner, the following described selections and optimization steps are normally done for the "Reference plant."

Compared to the beginning period of combined-cycle power plants, most of the Engineering, Procurement and Construction (EPC) companies, having their own gas turbine (product), applied a standardization approach based on actual market conditions and or needs. The development from tailor-made plants to standardized plants is shown in figure 5–2.

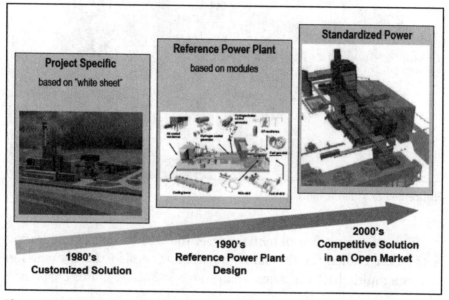

Figure 5–2 Evolutionary change in combined-cycle design philosophy

For the power island, the idea of this approach (with some exchangeable modules such as two or three options for the cold end of the steam turbine with condensation unit) is that the pre-engineered solution is fully reusable, and has the following benefits:

- Time (shorter delivery time)
- Quality (robust design)
- Risk (exchangeable components in case of problems)
- Cost (benefits by learning- and volume effect)

These benefits outweigh the small loss of efficiency compared to a fully customized designed power plant (see figure 5–3).

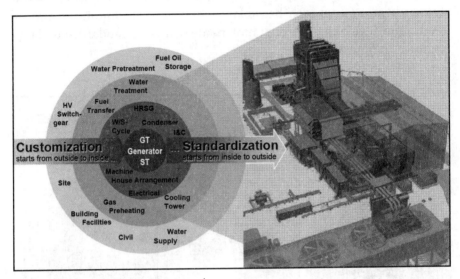

Figure 5–3 Standardization approach

With reference to a potential reuse, the one-to-one configuration is the best choice for the standardization. See also the comparison between multi-shaft and single shaft (table 5–3) later in this chapter.

Analysis of Customer Requirements

The customer (market) requirements define what a plant must be able to achieve in terms of performance and operational capability. The most significant requirement is the electricity power demand. When identified, it is important to determine any limits above or below the nominal power rating, which may be imposed, for example, by grid limitations or the need to meet an internal industry power demand. The best fit of turbines corresponding to this range must be found, taking into consideration the base load point at which the gas turbines will be operating.

If there is a steam-process demand it is necessary to look at the temperature, pressure, and mass flow required at the supply limit. The fluctuations allowed under these conditions and the demand variation over time should be known to design the plant for all possible operating conditions. If there is a return of steam or condensate from the industrial process, its condition and quality need to be considered in the cycle design as well (see chapter 6).

Combined-cycle power plants are intended for various uses such as base load, cycling, or frequency control. To account for such factors at the design stage, it is important to have information about expected operating philosophy, operating hours as well as number of starts per year, load scheme, grid requirements, and so on.

In case of a liberalized electricity market, financing becomes more and more important. This is especially true in the case of project financing. To be worthy of credit, any new power plant must be able to compete with other power plants on the open market on the basis of $/MWh generated, not just for the first year but throughout the pay-back period of the plant and preferably even longer.

Site-Related Factors

Site-related factors are specific to the intended location of the combined-cycle plant and are usually out of control of the power plant developer. Those that affect the choice of the cycle must be considered in the cycle selection process. One fundamental factor, which is usually country dependent, is whether the local electricity grid is rated at 50 or 60 Hz. This affects the selection of the gas turbine type because larger gas turbines are designed for specific frequencies. Smaller gas turbines are usually geared and, with an adequate gearbox, can operate with both of these frequencies.

Ambient conditions

Ambient conditions affect the performance of the cycle. This section looks at possible influences of these conditions regarding changes of the design point. How an already dimensioned combined-cycle plant behaves with different ambient conditions will be discussed in chapter 9.

The gas turbine is a standardized machine, and can be used for widely different ambient conditions. This can be justified economically because a gas turbine that has been optimized for an air temperature of 15°C (59°F) does not look significantly different from one that has been designed for air temperatures such as 40°C (104°F). In such a case the costs for developing a new machine would not be justified. Manufacturers quote gas turbine performances at ISO ambient conditions of 15°C (59°F), 1.013 bar (14.7 psia), and 60% relative humidity. The gas turbine will perform differently at other ambient conditions, and this will have an effect on the water/steam cycle process.

Unlike the gas turbine, the steam turbine is usually designed for a specific application. The exhaust steam section design, for example, depends on the condenser pressure at the design point, such as the exhaust section that would be chosen for a condenser design pressure of 0.2 bar (5.9"Hg). It can no longer function optimally if the operation pressure is only 0.045 bar (1.3"Hg). Also, blade path design in a steam turbine depends on the live-steam pressure, which is not the same for all cycles.

Nevertheless, as already explained, this detailed optimization and fine-tuning is done in most of the cases for the reference plant design and not anymore on a case-by-case basis, or for each and every project.

Air temperature and pressure are the most crucial ambient conditions. Relative humidity has only a minor influence, but becomes more important if the cooling concept is equipped with a wet cell (evaporative) cooling tower.

Ambient air temperature

There are three reasons why air temperature has a significant influence on the power output and efficiency of the gas turbine:

- Increasing the ambient temperature reduces the density of the air and consequently reduces the air mass flow into the compressor as constant volume engine. This is the main reason for changes in the gas turbine power output.

- The specific power consumed by the compressor increases proportionally to the air intake temperature (in K) without a corresponding increase in the output from the turbine part.

- As the air temperature rises and the mass flow decreases, the pressure ratio within the turbine is reduced. Due to the swallowing capacity of the turbine section and the reduced mass flow, the pressure at the turbine inlet of the GT is reduced. This leads to a lower pressure ratio within the turbine, and applies inversely, of course, to the compressor; however, because its output is less than that of the turbine, the total balance is negative.

Figure 5–4 shows the gas turbine characteristics at two different ambient pressures in a temperature/entropy diagram. The exhaust gas temperature rises as the air temperature increases because the turbine pressure ratio is reduced, although the gas turbine inlet temperature (TIT) remains constant. The result is a decrease in the gas turbine efficiency and output while the ambient temperature rises. However, the effect on the performance of the combined cycle is somewhat less because a higher exhaust gas temperature improves the performance (power output) of the steam cycle.

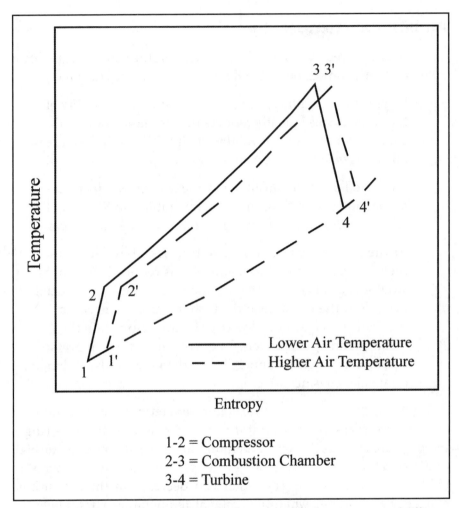

Figure 5–4 Entropy/temperature diagram for a gas turbine process at two different ambient air temperatures

Figure 5–5 shows the relative efficiency of the gas turbine, steam process, and the combined-cycle plant due to the air temperature; other ambient conditions as well as the condenser pressure remain unchanged. An increase in the air temperature has a slightly positive effect on the efficiency of the combined-cycle plant. Because the increased temperature in the gas turbine exhaust enhances the efficiency of the steam process, it more than compensates for the reduced efficiency of the gas turbine unit.

This behavior is not surprising considering the Carnot efficiency (equation 4–1). The increase in the compressor outlet temperature causes a slight increase in the average temperature of the heat supplied (T_E). Most of the exhaust heat is dissipated in the condenser; therefore, the cold temperature (T_A) does not change because the condenser pressure is kept constant. The overall efficiency of the combined-cycle plant will increase.

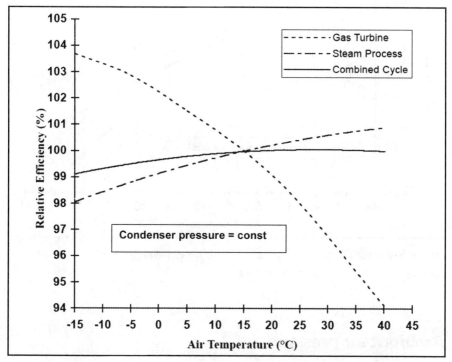

Figure 5–5 Relative efficiency of gas turbine, steam process, and combined cycle as function of the air temperature

Figure 5–6 shows how the power output of the gas turbine, steam turbine, and combined cycle decrease with an increase in the air temperature. The effect is less pronounced for the combined cycle than for the gas turbine alone. The power output of the combined cycle is affected differently from the efficiency because change in mass flow of air and exhaust gases are more dominant than the exhaust gas temperature.

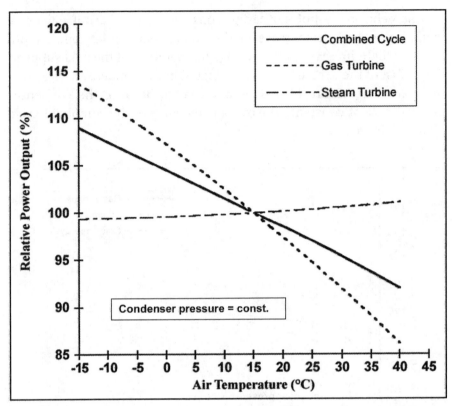

Figure 5–6 Relative power output of a gas turbine, steam turbine, and combined cycle as function of the air temperature

Ambient air pressure

Gas turbine performance is normally quoted at an air pressure of 1.013 bar (14.7 psia)—ISO condition, which corresponds approximately to the average pressure prevailing at sea level. A different site elevation and daily weather variations result in a different air pressure.

Figure 5–7 shows the relationship between site elevation and ambient air pressure, and its influence on relative power output of the gas turbine, steam turbine, and combined cycle. The air pressure has no effect on the efficiency if the ambient temperature is constant, even though the ambient air pressure has an influence on the air density similar to the one of the air temperature. At a lower ambient pressure, the backpressure of the gas turbine is correspondingly lower, not considering inlet and outlet pressure drops. This leaves no influence on the gas turbine process with the exception of the reduced mass flow. Assuming that no change takes place in the efficiency of the steam process (which corresponds well to the real situation), there is the same variation in the power output of the steam turbine and hence the combined cycle.

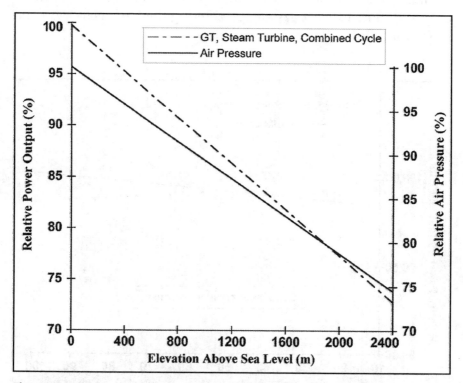

Figure 5–7 Relative power output of gas turbine, steam turbine, and combined cycle and relative air pressure versus elevation above sea level

Because the power outputs of the gas turbine and the steam turbine vary in proportion to the air pressure, total power output of the combined-cycle plant varies in proportion. The fact that the gas turbine inlet and outlet pressure drops were held constant for the calculations of figure 5–7 accounts for the slight difference in the relative power output compared to the air pressure. The efficiency of the plant remains constant because both the thermal energy supplied as well as the airflow vary in proportion to the air pressure.

Relative humidity

Figure 5–8 shows that the gas turbine and combined-cycle output will increase if the relative humidity of the ambient air increases, with other conditions remaining constant. This is because at higher levels of humidity there will be a higher water content in the working medium of the gas cycle, resulting in a better gas turbine enthalpy drop and more exhaust gas energy entering the heat recovery steam generator (HRSG).

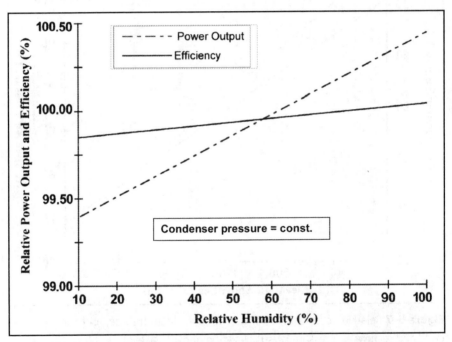

Figure 5–8 Relative power output and efficiency of gas turbine and combined cycle as function of relative humidity

There is a further influence for plants with wet cooling towers where the relative humidity directly influences the condenser vacuum, and hence the steam turbine exhaust temperature. A lower humidity results in a better vacuum. This is discussed in the "Resources" section later in this chapter.

Legislation

Many environmental considerations affect design and construction of all types of power stations. The choice of a combined-cycle concept is mainly influenced by legislation. This legislation limits emission levels, especially that of nitrous oxides (NO_x). Current dry low NO_x gas turbine combustors can often achieve the required levels with gaseous fuel, but special measures may have to be taken if oil fuels are fired.

One option to reduce the formation of NO_x during combustion is to lower the temperature of the flame because the speed of the reaction producing NO_x is noticeably faster at very high temperatures. The common approach for gaseous fuels is to use premix burners to lower the flame temperature, mixing air at the outlet of the compressor with the fuel before it is ignited in a vortex breakdown zone. Injecting water or steam into the combustor can produce the desired temperature reduction; however, it will affect the output and efficiency of the gas turbine. This approach is, therefore, mainly used for fuel oil where premix is problematic due to the risk of pre-ignition. The amount of possible NO_x reduction depends on the water or steam to fuel ratio as described in chapter 10.

For gas turbines in simple cycle with no HRSG (and, therefore, no available steam source), water is injected for NO_x control, but with a negative effect on efficiency.

The problem with steam injection is finding steam at a suitable pressure level in the steam cycle. Generally the high-pressure (HP) live-steam pressure is too high and the intermediate-pressure (IP), in case of GT's with a high compressor ratio and low-pressure (LP) steam pressure is too low. For large industrial gas turbines the pressure level required for steam injection is between 25 and 50 bar (348 and 710 psig), depending on the type of gas turbine as well as the load. Using pressure-reduced HP steam is usually the simplest and the least expensive solution, but is exergetically undesirable.

For this reason, a base-load plant with steam injection into the GT will usually have a steam extraction in the steam turbine at the appropriate pressure level, with a backup from the HP live-steam line for those operating points at where the extraction pressure is too low. Steam extraction is, however, more problematic with steam turbines in combined cycle blocks with several gas turbines. Unless all the gas turbines are in operation, the pressure at the extraction point decreases so far that it is, in most cases, inadequate. Thus it becomes necessary to switch over to live steam, which negatively affects the overall efficiency.

Power output is increased by both water and steam injection due to the resulting increase in the mass flow through the gas turbine. Water injection has a greater effect because steam turbine output is decreased with steam injection. Water or steam injection is also a means of power augmentation (see also chapters 9 and 6). Efficiency of the combined-cycle plant is decreased in both cases; however, less so by steam than by water, because steam brings more internal energy to the combustor. In the case of water, valuable fuel has to be used to provide the relatively low temperature energy for the vaporization. For the same reason hotter water partly reduces the loss in efficiency.

Figure 5–9 shows the effect that injecting water or steam has on the output and efficiency of a dual-pressure cycle. The steam is extracted from the steam turbine, the water from the makeup water line at 15°C (59°F) or at the outlet of the economizer at 150°C (302°F). The amount of water or steam is provided by the water or steam to fuel ratio, which is determined by the necessary reduction level of NO_x (or, in very few special cases, by the amount of power augmentation required).

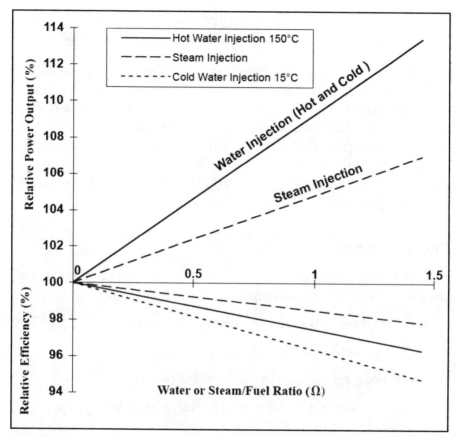

Figure 5–9 Effect of water and steam injection on relative combined-cycle power output and efficiency versus water or steam/fuel ratio (with TIT = constant)

Apart from NO_x, other emissions are often restricted, such as CO, particulates, and so forth. Those emissions influence the component design rather than the cycle performance, however, and so do not influence the selection of the concept.

Another restriction could be a limitation of the cooling water temperature in case of fresh-water cooling (with sea water or river water) either by a limitation of a maximum allowable temperature rise, or with a limitation of the maximum allowed cooling water outlet temperature or restrictions regarding visible plumes in case of a wet cooling tower. This is discussed in the following section.

Resources

Available resources are different at every site. Cooling media and fuel are the main resources affecting the selection of the cycle concept. Although space is also a resource, it affects the cycle only when space is restricted by, for example, the need to use existing buildings or part of an existing cycle. This may lower the capital cost of the plant, but could also limit the possibilities for performance improvements by, for example, restricting the size of the HRSG or influencing the choice of shaft configuration.

Cooling media

To condense the steam, a cooling medium must be used to carry off the waste heat from the condenser.

There are three options for the main cooling system:

- Fresh-water cooling (with sea or river water)
- Evaporation cooling (wet cell cooling tower)
- Dry cooling system (in case of a direct system with an air-cooled condenser [ACC])

Cooling with water has a clear advantage because water has a high specific thermal capacity and good heat transfer properties. Therefore, where water is available in required quantities, cooling can be done in a direct system (fresh water from a river or the sea) or in a wet cooling tower. The cooling water quantities in case of fresh water cooling are approximately 60 to 100 times the exhaust steam flow (in some special cases with very limited temperature rise even higher) and for wet cooling systems as cooling tower makeup water around 1 to 1.4 times the exhaust steam flow or approximately a minimum amount of 0.3 kg/s (2400 lb/h) per MW installed combined-cycle capacity (without supplementary firing).

To ease the permitting process where water is not available, a direct air-cooling system with an air-cooled condenser (or a dry cooling tower as indirect system) is used. The disadvantages of this method include higher costs and higher exhaust steam (condenser) pressure, mainly in the off-design points with high ambient temperatures (therefore less power output of the steam turbine) leading to lower plant efficiency. In combined-cycle plants with a direct air-cooling system (ACC) the ambient air temperature directly influences the ST output and consequently the efficiency, which is not (or to a less extent) the case for the other cooling system (see table 5–1).

Table 5–1 Comparison of combined cycle performance data for different cooling systems and ambient temperatures

	Design: ambient temperature = 6°C (43°F)		Off-design: Summer ambient temp = 30°C (88°F)	
	P MW	Efficiency η %	P MW	η%
Cooling tower	411.7 100%	57.5 100%	363.5 100%	56.1 100%
ACC	408.1 99.1%	57.0 99.1%	345.4 95%	53.3 95%

In special cases where only a limited amount of water is available, and an evaporation system could be used with restrictions regarding visible plumes, a combination of a dry and wet system with hybrid cooler cells (consisting of a dry and a wet section) could be the optimal solution. Design parameters/performance data are similar to the wet cell cooling tower, but would result in slightly higher costs.

The temperature of the cooling medium has a major effect on the efficiency of the thermal process. The lower the temperature the higher the efficiency that can be attained, because the pressure in the condenser is lower, producing a greater useful enthalpy drop in the steam turbine and hence an increase in steam turbine output and in plant efficiency. This is illustrated in figure 5–10.

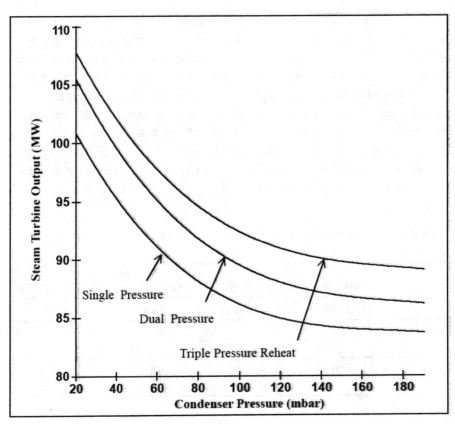

Figure 5-10 Effect of condenser pressure on steam turbine output

The trend is similar for single-, dual-, and triple-pressure cycles (discussed later in this chapter). However, the effect is less significant above approximately 100 mbar (3.0"Hg) condenser pressure as the relative change in the pressure decreases and there is less impact on the steam turbine enthalpy drop. Plant costs are reduced if the pressure is higher because of the lower volume flow of exhaust steam resulting in a smaller steam turbine. This impact of a change in the condenser vacuum is increased with a dual- or even triple-pressure than with a single-pressure cycle because the exhaust steam flow is larger.

Figure 5–11 shows typical condenser design pressure values as function of the temperature of the cooling medium for direct water cooling (fresh water), cooling with a wet cooling tower, and direct air cooling. These values are calculated based on fuel gas price levels (2007)

in Europe and the United States. The best vacuums are attained with direct water cooling; the worst ones are with direct condensation with air, mainly with ambient air temperatures higher than 20 to 25°C (68 to 77°F). In comparison with ambient air temperature, the corresponding water temperature is generally lower. The curves for direct cooling and cooling tower cooling are calculated with cooling water temperature rises of 9K and 11K, respectively.

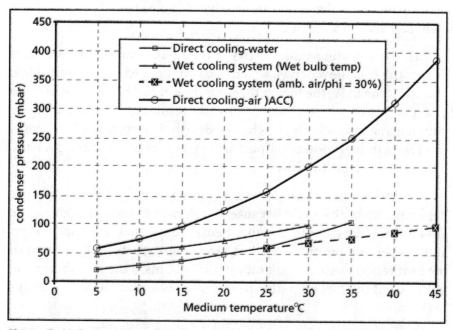

Figure 5–11 Temperature of cooling medium versus condenser pressure for different types of cooling systems

In case of wet cell cooling tower, the wet bulb temperature is determined as temperature of the cooling medium. It is a function of the dry bulb temperature and the relative humidity, and can be read from an enthalpy-moisture content diagram for air (Mollier diagram). Out of the statistical values the maximum is not going beyond about 30°C (86°F). Therefore, in case of higher ambient air temperatures in the range of about 30 to 45°C, a relative humidity (phi) of 30% was selected in figure 5–11 for comparison reasons.

The influence on the steam turbine output of the various cooling systems can be derived out of figure 5–10 in combination with figure 5–11.

Fuel

This section explains how the sulfur content in the fuel influences the cycle concept because of the feedwater temperature. The fuel type and composition also have a direct influence on the gas turbine performance and the emissions produced. Power plants often have a main fuel and a backup fuel.

The lower heating value (LHV) of the fuel is important because it defines the mass flow of fuel, which must be supplied to the gas turbine. The lower the LHV, the higher the mass flow of fuel required to provide a certain chemical heat input, normally resulting in a higher power output and efficiency. This is why low British thermal unit (BTU) gases can result in high-power outputs if they are supplied at the pressure required by the gas turbine.

The fuel composition is equally important in influencing the performance of the cycle because it determines the enthalpy of the gas entering the gas turbine and, hence, the available enthalpy drop and gas turbine output as well as the amount of steam generated in the heat recovery steam generator (HRSG). This is the reason why the influence of the fuel on performance cannot be given as a function of LHV only.

One way of improving the efficiency of the cycle is to raise the apparent LHV (LHV + sensible heat) of the fuel by preheating it with hot water from the economizer of the HRSG. This improves heat utilization in the HRSG because additional water is heated in the economizer. The fuel consumption is correspondingly reduced because each unit of fuel contains more sensible heat, so a lower fuel mass flow is required. Figure 5–12 shows schematically a hot water extraction from the economizer (5) of a dual-pressure HRSG. The cooled water (4) returns to the feedwater tank. Preheating natural gas fuel from 15°C (59°F) to 150°C (302°F) can improve the plant efficiency by approximately 0.7% (relative). Today, depending on the type of gas turbine, fuel gas preheating can go up to 250°C (482°F); if hot water is not available, it can also be done economically with steam.

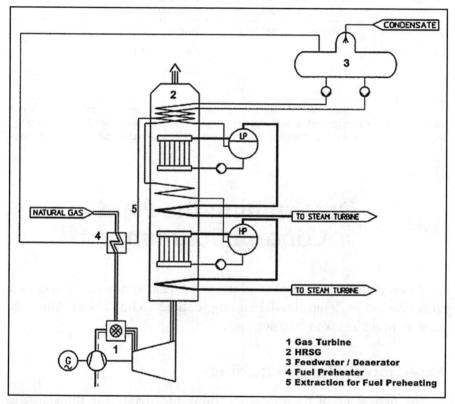

Figure 5–12 Flow diagram to show fuel preheating

Another factor, which may be important for gas-fired plants, is the available gas supply pressure. The gas turbine requires a certain pressure, depending on the design of the burners in the combustor and the gas turbine pressure ratio. Sometimes a gas compressor is required to increase supply pressure, which will also increase the temperature of the fuel in proportion to the compression ratio. The benefit of gas preheater in such cases is reduced, perhaps by even rendering it uneconomical because the efficiency improvement is too slight to justify the additional investment in the water/gas heat exchanger, HRSG surface, and piping. (See chapter 7 for more information about fuels for gas turbines.)

Table 5–2 lists the fuels that can be fired in a gas turbine.

Table 5–2 Possible fuels for combined cycle applications

a) Standard Fuels	b) Special Liquid fuels	c) Special Gas Fuels
Natural gas	Methanol/Kerosene/Naphtha	Synthetic gas
Diesel oil (oil # 2)	Crude oil	Blast furnace gas
	Heavy oil, residuals	Coal gas with medium or low heating value
	Oil shale	

Note: The use of categories b) and c) is limited, because their burning depends on the exact chemical analysis and the type of gas turbine involved. Generally, industrial gas turbines with large combustors are better able to handle these fuels than modern gas turbines with high turbine inlet temperatures.

Determining the Plant Concept Solution

When plant requirements and the site data are known, a concept for the cycle can be determined. In doing so, both technical and economic aspects must be taken into account.

Selection of the gas turbine

The first stage of the cycle selection is to determine the size and number of gas turbines needed to meet a certain power output and eventually a certain process steam production.

For a plant with a given gas turbine, supplying process steam will decrease the power output because steam is removed from the steam cycle rather than being used for power generation. As a result, a larger gas turbine could be selected, rather than if process steam were not required. A gas turbine of a lower rating could be chosen for a given application, with peak power produced by an additional steam turbine output through supplementary firing (or with other power augmentation methods). With regard to the steam turbine, it is of no importance if the required exhaust gas temperature is attained directly from the gas turbine exhaust or by means of supplementary firing in the HRSG. However, from an efficiency point of view, preference would be to a fuel feeding directly in the gas turbine at a higher exergetic level than in the HRSG duct burner.

If more than one gas turbine is needed, a choice must be made between a multi-shaft configuration or several single shaft blocks. Several manufacturers have developed single shaft power plants (power islands) for their gas turbines. These standardized units provide advantages related to fast installation times, availability, and space requirements. Table 5–3 shows the comparison between a 2–1 multi-shaft combined cycle and two single-shaft units.

Table 5–3 Comparison multi-shaft versus single shaft (2 single shaft units versus a multi shaft 2-1 unit)

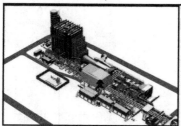

Components	Single shaft:	Less generators required Less HV-switchyards required Less electrical equipment required One compact lube oil system
	Multi shaft:	No synchronous clutch required One large versus 2 smaller ST Less auxiliaries (pumps etc) required
Civil	Single shaft:	Smaller footprint required
	Multi shaft:	Higher flexibility in plant layout
Costs	Single shaft:	Lower specific costs of plant if standardized Lower cost for grid connection
Performance	Single shaft:	At full load both solutions have almost the same performance level in case of larger plants
Flexibility	Multi shaft:	Faster start-up time from half load operation to full load
Availability	Single shaft:	Higher availability (less complexity – clear maintenance concept)

Not all gas turbines with the same inlet airflow have the same performance. It depends on the turbine inlet temperature (TIT), the design concept (i.e., one or two stages of combustion), and the gas turbine exhaust gas temperature (which is a function of the pressure ratio). Because this has a marked influence on the performance of the combined cycle, it is worth looking at in more detail.

When the gas turbine exhaust temperature is lowered, both the thermodynamic quality of the steam process (seen by the steam turbine output) and the HRSG efficiency decrease as shown in figure 5–13.

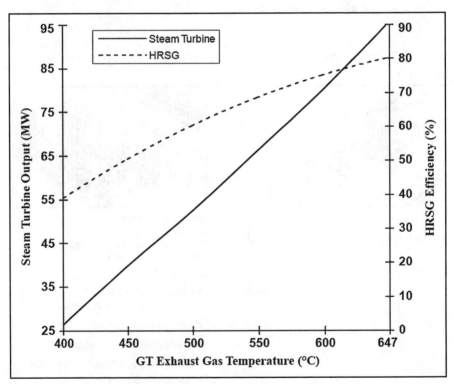

Figure 5–13 Steam turbine output and HRSG efficiency versus gas turbine exhaust temperature for a single-pressure cycle

The effect is more pronounced with the single pressure than with a dual or even triple-pressure reheat cycle because the energy utilization rate falls off faster.

Figure 5–14 shows the ratio between the output of the dual-pressure and the single-pressure cycles as a function of the gas turbine exhaust gas temperatures.

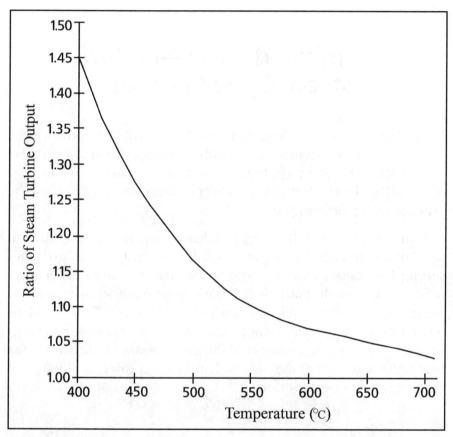

Figure 5–14 Ratio of steam turbine output of a dual-pressure compared to a single-pressure cycle as a function of the gas turbine exhaust temperature

With increasing exhaust gas temperature, the advantage of a dual-pressure versus a single-pressure cycle diminishes more and more, and at a certain temperature the ratio equals practically to 1. The theoretical temperature limit is a function of the chosen steam parameters (temperatures and pressures). For normal parameters this limit can be reached at around 750°C. For highly advanced and individually optimized cycles, this limit can theoretically slide up to around 900°C or higher. Because cycles with (high) supplementary firing are mostly derived from normal cycles (e.g., single or dual-pressure cycles), the lower limit will apply here. Therefore, cycles with (high) supplementary firing are mostly single-pressure cycle.

Combined Cycle—Water/Steam Cycle Concepts

The main challenge in designing a combined-cycle plant with a given gas turbine is how to transfer the gas turbine exhaust heat to the water/steam cycle to achieve an optimum steam turbine output. The focus is on the HRSG in which the heat transfer between the gas cycle and the water/steam cycle takes place.

Figure 5–15 shows the energy exchange that would take place in an idealized heat exchanger, in which the product (mass flow times specific heat capacity, or the energy transferred per unit temperature) must be the same in both media at any given point so that there are minimum exergetic losses in a given heat exchanger. There must be a temperature difference between the two media to allow the energy transfer. Because the temperature difference tends towards zero, the heat transfer surface of the heat exchanger tends towards infinity and the exergy losses towards zero.

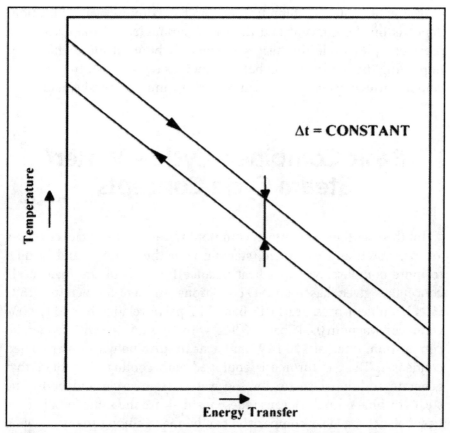

Figure 5–15 Energy/temperature diagram for an idealized heat exchanger

The heat transfer in a HRSG entails losses associated with three main factors:

- The physical properties the of water, steam, and exhaust gases do not match the higher exergetic losses than the ones of an idealized heat exchanger as shown in figure 5–15.
- The heat transfer surface cannot be infinitely large.
- The temperature of the feedwater must be high enough to prevent forming corrosive acids in the exhaust gas, where it comes in contact with the cold tubes in the HRSG. This minimum feedwater temperature limits the energy utilization in the HRSG (temperature to which the exhaust gas can be cooled).

The extent of loss minimization (and maximum heat utilization) depends on the concept and the main parameters of the cycle. In a more complex cycle the heat will generally be used more efficiently, improving the performance, but also increasing the cost. In practice a compromise between performance and cost must always be made.

Basic Combined-Cycle—Water/Steam Cycle Concepts

In this section, the most common water/steam cycle concepts are presented and explained, starting with the simplest and leading to more complex cycles. A heat balance for each of the main cycle concepts is given, based on ISO conditions (ambient temperature 15°C (59°F); ambient pressure 1.013 bar, (14.7 psia); relative humidity 60%; condenser vacuum 0.045 bar (1.3"Hg) with a gas turbine with sequential combustion, rated at 178 MW and a steam turbine with water cooled condenser. The gas turbine is equipped with cooling air coolers that generate additional steam for the water/steam cycle and boost the steam turbine output. Because these features are the same for all of the heat balances, a clear comparison can be made among them, showing how the cycle concept influences the heat utilization.

Additionally, an analysis is done for each concept to show how the main cycle parameters, such as live-steam temperatures and pressures, and HRSG design parameters, influence the performance of the cycle. When designing a combined cycle with a given gas turbine, the free parameters are those of the steam cycle. Therefore, the influence of the various parameters is analyzed with respect to the steam turbine output. This is because in a plant without additional firing, the thermal energy supplied to the steam process is given by the gas turbine exhaust gas and the efficiency of the steam process is always proportional to the steam turbine output. The steam turbine does, however, account for only about 30 to 40% of the total combined-cycle power output; therefore, optimization of the steam process can only influence that portion.

Although the cycles shown have only one gas turbine, the concepts and results are also valid for cycles with several gas turbines, HRSG's, and other types of industrial gas turbines.

Single-pressure cycle

The simplest type of combined cycle is a basic single-pressure cycle, so called because the HRSG generates steam for the steam turbine at only one pressure level. A typical flow diagram is shown in figure 5–16 with a gas turbine exhausting into a single HRSG.

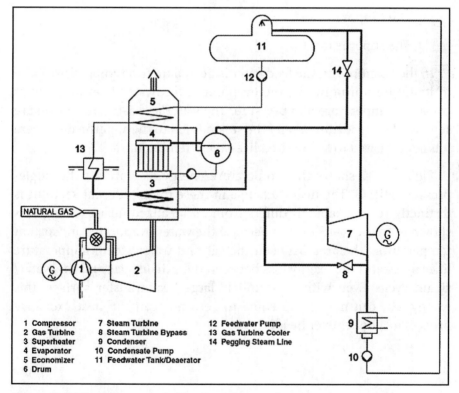

1 Compressor
2 Gas Turbine
3 Superheater
4 Evaporator
5 Economizer
6 Drum
7 Steam Turbine
8 Steam Turbine Bypass
9 Condenser
10 Condensate Pump
11 Feedwater Tank/Deaerator
12 Feedwater Pump
13 Gas Turbine Cooler
14 Pegging Steam Line

Figure 5–16 Flow diagram of a single-pressure cycle

The steam turbine (7) has a steam turbine bypass (8) into the condenser, which is used to accommodate the steam if for any reason it cannot be admitted to the steam turbine, for example during startup, in case of a steam turbine trip, or because the steam turbine is out of operation. After the condenser (9), a condensate pump (10) is used to pump the condensate back to the feedwater tank/deaerator (11). The feedwater pump (12) returns the feedwater to the HRSG. Heating steam for the deaerator is extracted from the steam turbine with a pegging steam supply from the HRSG drum, in case the pressure at the steam turbine bleed point becomes too low at off-design conditions.

The HRSG consists of three heat exchanger sections:

- The economizer (5)
- The evaporator (4)
- The superheater (3)

In the economizer, the feedwater is heated up to a temperature close to its saturation point. In the evaporator, feedwater is evaporated at constant temperature and pressure. The water and saturated steam are separated in the drum (6) and the steam is fed to the superheater where it is superheated to the desired live-steam temperature.

Figure 5–17 shows the temperature/energy diagram for the single-pressure HRSG. The heat exchange in these three different sections is distinctly recognizable. It differs from an idealized heat exchanger as shown in figure 5–15 mainly because the water evaporates at constant temperature. The area between the gas and water/steam temperature lines illustrates the exergy loss between the exhaust gas and the water/steam cycle. Even with an infinitely large heat transfer surface, this exergy loss can neither be equal to zero nor can the heat exchange process in a boiler ever be ideal.

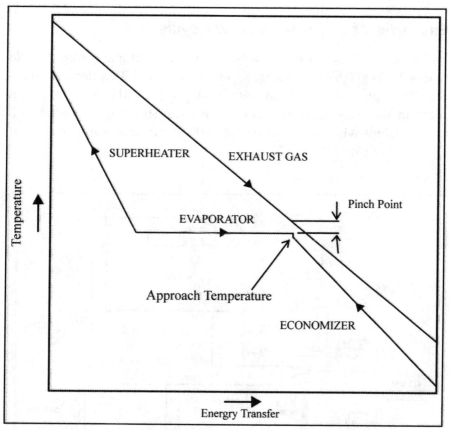

Figure 5–17 Energy/temperature diagram of a single-pressure HRSG

Two important parameters defining the HRSG are marked on the diagram in figure 5–17. The approach temperature is the difference between the saturation temperature in the drum and the water temperature at the economizer outlet. This difference, typically 5 to 12K (9 to 22°F), helps to avoid evaporation in the economizer at off-design conditions.

The pinch point temperature is the difference between the evaporator temperature on the water/steam side and the outlet temperature on the exhaust gas side. This is important when defining the heating surface of the HRSG. The lower the pinch point, the more heating surface is required, and the more steam is generated. Pinch points are typically between 8 and 15K (14 to 27°F), depending on the economic parameters of the plant.

Example of a single-pressure cycle

Figure 5–18 shows the heat balance for the single-pressure cycle where 73.3 kg/s (579'400 lb/hr) steam at 105 bar (1508 psig) and 568°C (1054°F) is generated. A small loss of temperature and pressure is taking place in the live-steam line, after which the steam is expanded in a steam turbine with an output of 94.8 MW. The resulting gross electrical efficiency of this cycle is 57.7%.

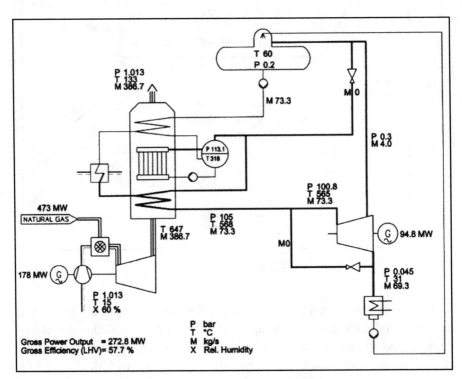

Figure 5–18 Heat balance for a single-pressure cycle

Energetic utilization of the exhaust heat is relatively low, considering that the feedwater temperature is 60°C (140°F) as illustrated by the stack temperature of 133°C (271°F), as figure 5–19 shows in an energy flow diagram; 11.4% of the fuel energy supplied is lost through the boiler stack, another 29.9% is discharged in the condenser, and about 1% is equipment losses.

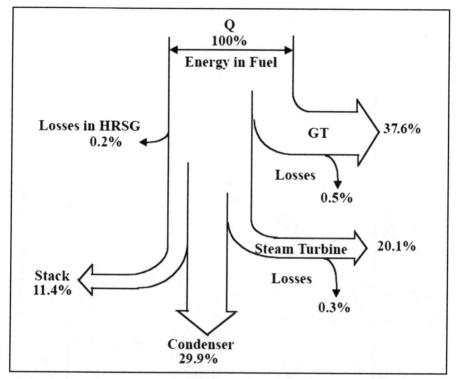

Figure 5–19 Energy flow diagram for the single-pressure combined-cycle power plant

Main design parameters of the single-pressure cycle

Live-steam pressure. In a combined-cycle plant, a high live-steam pressure does not necessarily mean a high efficiency of the cycle. Figure 5–20 shows how the steam turbine output varies with live-steam pressure for a single-pressure cycle. Expanding the steam at a higher live-steam pressure will give a higher steam turbine output due to the greater enthalpy drop in the steam turbine. However, because of the higher evaporation temperature in the HRSG, less steam will be generated, resulting in a higher exhaust gas temperature and, therefore, in a lower HRSG efficiency. An optimum is normally found between these two influences. In the pressure range shown in fig. 5–20, the optimum live-steam pressure is actually higher than 110 bar. However, at 110 bar the moisture content in the steam exhausted into the condenser is already 16%, which is actually quite high for the last stage of the steam turbine and is, therefore, the limiting factor for the selection of the live-steam pressure.

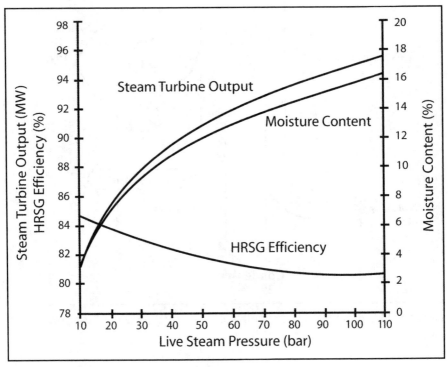

Figure 5–20 Effect of live-steam pressure on steam turbine output for a single-pressure cycle (including steam turbine exhaust moisture content and HRSG efficiency)

The impact of the moisture content on the polytropic steam turbine efficiency is shown in the appendices (equation A–29).

It is interesting that the high HRSG efficiency does not correspond to the high steam turbine output. At a lower live-steam pressure there is a lower stack temperature and the energy in the exhaust gas is well utilized. However, the steam turbine output at this live-steam pressure is lower because of higher exergy losses in the HRSG. The reverse happens at higher dry air cooling section. It shows that exergy is more dominant than energy in determining the steam turbine output.

This effect is illustrated in figure 5–21, the energy/temperature diagram for two different single-pressure cycles with live-steam pressures of 40 and 105 bar (566 and 1508 psig). At the lower live-steam pressure, more thermal energy is available for evaporation and superheating because the evaporation temperature is correspondingly lower. The pinch point of the evaporator is the same in both cases. As a result, the stack temperature at 40 bar is about 11°C (20°F) lower than at 105 bar, which means that more waste heat energy is being utilized.

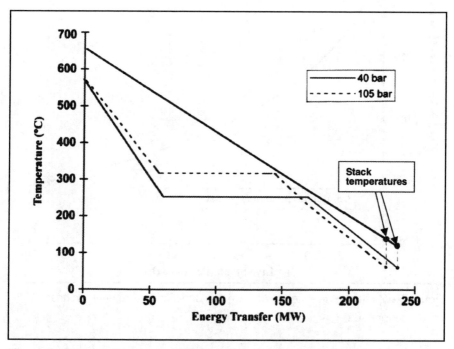

Figure 5–21 Energy/temperature diagram of a single-pressure HRSG with live-steam pressure of 40 and 105 bar (566 and 1508 psig)

A negative aspect of a higher live-steam pressure in a single-pressure cycle is the increase in the moisture content at the end of the steam turbine. Too much moisture increases the risk of erosion, taking place in the last stages of the turbine. A limit is set at about 16%. The moisture content depends also very much on the condenser pressure: the higher the pressure, the lower the moisture content.

A change in the live-steam pressure affects the amount of heat to be removed in the condenser as shown in figure 5–22 (because of change in the steam mass flow).

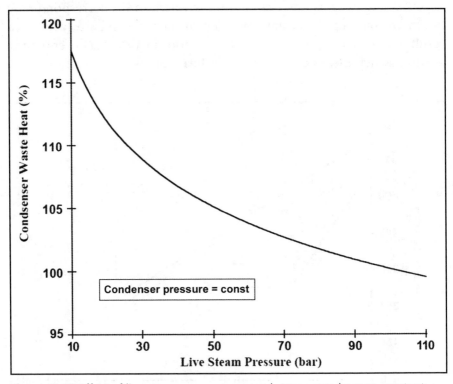

Figure 5–22 Effect of live-steam pressure on condenser waste heat at constant condenser pressure

A higher live-steam pressure creates the following economical advantages:

- A smaller exhaust section in the steam turbine
- A smaller volume flow leading to smaller live-steam piping and valve dimensions
- A reduction of the cooling water requirement
- Smaller cooling water equipment (condenser, pumps, cooling tower, etc.)

This can lead to considerably lower costs, especially for power plants with expensive air-cooled condensers.

The total amount of live steam can also influence the optimum live-steam pressure because it has an influence on the steam turbine efficiency. Increasing the amount of live steam will result in a larger volume flow at the turbine inlet, leading to longer blades in the first row of the turbine, which reduces the secondary blading losses. It follows that the optimum live-steam pressure also depends on the type of gas turbine used because different exhaust gas conditions will result in different live-steam flows for a given live-steam pressure.

In the example, a pressure of 105 bar (1508 psig) is chosen because it is the highest pressure possible without excessive moisture content in the steam turbine exhaust.

Live-steam temperature. For the chosen live-steam pressure as shown in figure 5–23, raising the live-steam temperature causes a slight decrease in steam turbine output. This is the result of two opposite effects. Increasing the live-steam temperature (same as when live-steam pressure is increased) results in a greater enthalpy drop in the steam turbine; however, at the same time, additional superheating removes energy, which would otherwise be used for steam production, resulting in a lower steam flow as well as a corresponding loss in the steam turbine output and an increase in the stack temperature. The latter is the dominant effect in this case.

The live-steam temperature cannot be reduced below a certain limit for a given live-steam pressure because of the resulting increase in the moisture content in the steam turbine. To use a lower live-steam temperature, the pressure would have to be reduced, which, would result in a more negative effect on the steam turbine output than raising the live-steam temperature.

In the example with 105 bar (1508 psig) live-steam pressure, a temperature of 568°C (1054°F) has been chosen, which is close to the upper limit for current materials used for these plants.

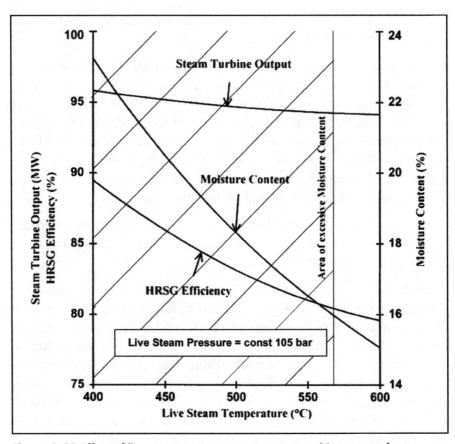

Figure 5–23 Effect of live-steam temperature on steam turbine output for a single-pressure cycle with 105 bar (1508 psig) live-steam pressure (including HRSG efficiency and steam turbine exhaust moisture content)

For gas turbines with lower exhaust gas temperatures than the one in the example, a lower live-steam pressure level would have to be chosen for the reasons previously discussed.

Design Parameters of the HRSG

Pinch point and approach temperature

An important parameter in the optimization of a steam cycle is the pinch point of the HRSG (defined in figure 5–18), which directly affects the amount of steam generated. Figure 5–24 shows how by reducing the pinch point (with other parameters kept constant) steam turbine output can be increased. This is due to a better rate of heat utilization in the HRSG. However, the surface of the heat exchanger—and, hence, the cost of the HRSG – increases exponentially as the pinch point tends towards zero. In figure 5–24, a reference pinch point of 12K (21.6°F) is used.

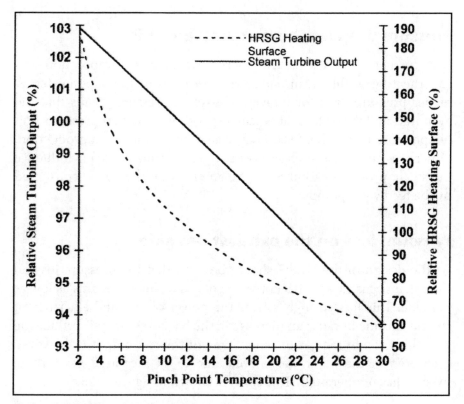

Figure 5–24 Effect of pinch point on relative steam turbine power output and relative HRSG heating surface

Similarly, a smaller approach temperature (defined in figure 5–18) leads to a better heat utilization but increased HRSG surface. For a drum-type HRSG, a lower limit is set on the approach temperature by the need to minimize steaming in the economizers for off-design points. Our examples use 5K (9°R) for the approach temperature.

It is the sum of the pinch point and approach temperatures that determine the steam production in the HRSG for a given live-steam pressure and temperature. This means that a HRSG with 10K (18°F) pinch point and 5K (9°F) approach temperature would have the same steam production as one with 5K (9°F) pinch point and 10K (18°F) approach temperature. However, the HRSG surface will not be the same in each case (due to a different logarithmic temperature difference in heat exchange) and so the optimum must be found, bearing in mind the limitation of the steaming in the economizers and the impact of the HRSG surface.

Pressure drop on the water/steam side

A further influence on the energy available for evaporation and superheating is the steam side pressure drop in the superheater. A higher pressure drop for a given live-steam pressure means that the evaporation takes place at a correspondingly higher pressure and temperature level, where less energy is available for steam production. Other pressure losses, such as those in the economizers, do not influence the steam production, but will have an effect on the power consumed by the feedwater pumps.

Pressure loss on the exhaust gas side

The design of the HRSG should be such that the pressure loss on the exhaust gas side (or backpressure of gas turbine) remains as low as possible. This loss strongly affects the power output and efficiency of the gas turbine because an increase in the backpressure will reduce the enthalpy drop in the gas turbine. Some of the lost output is, however, recovered in the water/steam cycle due to an increased gas turbine exhaust gas temperature. This effect is shown in figure 5–25.

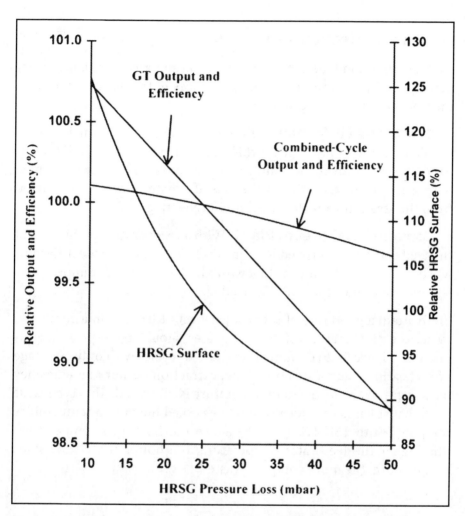

Figure 5–25 Influence of HRSG backpressure on combined-cycle output and efficiency, GT output and efficiency, and HRSG surface

Lower pressure losses on the flue gas side, however, have a negative impact on the heat transfer due to lower gas velocities around the tube bundle leading to higher HRSG surface. Typical HRSG exhaust gas pressure losses are between 25 to 30 mbar (9.9 to 11.8" WC). In the example, 25 mbar is used.

Feedwater preheating

Between a combined-cycle power plant and a conventional steam power plant, there are significant differences in the design/concept of the feedwater preheating system.

Combined-cycle plant. Based on equation (4–2) it is clearly visible that with a given gas turbine (power output and fuel input is fixed) the efficiency of the plant will be maximized if the steam turbine output is maximized. This is the case if no steam will be extracted from the steam turbine for feedwater preheating.

Conventional Steam Plant. With a regenerative preheating of the feedwater, an improved Rankine cycle (closer to the ideal Carnot) will be achieved. This means that with a higher feedwater temperature, less fuel input into the boiler is needed.

To attain a good rate of exhaust gas heat utilization in a combined cycle, the temperature of the feedwater should be kept as low as possible. Figure 5–26 demonstrates that for cycles with only one stage of preheating using a steam turbine, extraction output and efficiency fall sharply as the feedwater temperature is increased. This is because the exhaust gases can, ideally, only be cooled down to a temperature of about 10 to 15K (18 to 27°F) above the feedwater temperature. The higher the feedwater temperature, the hotter the HRSG stack temperature, the more energy is wasted.

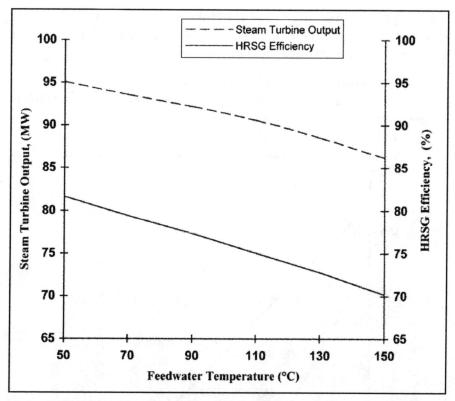

Figure 5–26 Effect of feedwater temperature on steam turbine output and HRSG efficiency for cycles with one stage of preheating

This is one significant difference between a conventional steam plant with a high feedwater temperature and the steam process in a combined-cycle plant. A conventional steam plant attains better efficiency if the temperature of the feedwater is brought to a high level by means of multistage preheating. There are two reasons for this difference.

A conventional steam generator is usually equipped with a regenerative air preheater that can further utilize the energy remaining in the flue gases after the economizer. This is not the case in a HRSG, where the energy remaining in the exhaust gases after the economizer is lost. In principle, air coming into the gas turbine could be similarly preheated, further lowering the stack temperature. However, this would lower the plant output significantly due to the decrease in air density at the compressor inlet and the fact that the gas turbines are volumetric machines, so that the inlet air volume flow to the gas turbine is always constant, however the mass flow decreases.

Also, as shown in figure 5–27, the smallest temperature difference between the water and exhaust gases in the economizer of an HRSG is on the warmer end of the heat exchanger.

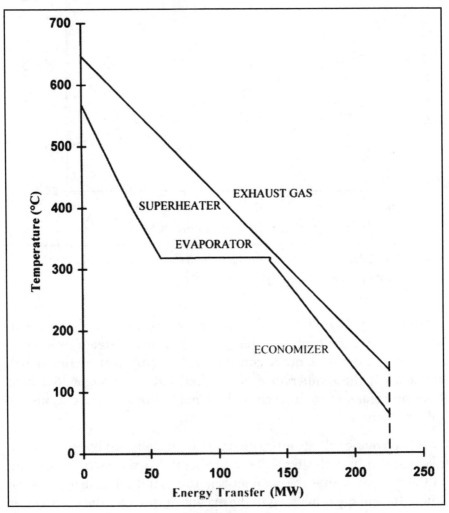

Figure 5–27 Energy/temperature diagram for a single-pressure HRSG

That means the amount of steam production possible does not depend on the feedwater temperature. In a conventional steam generator, on the other hand, the smallest temperature difference is on the cold end of the economizer because the water flow is far larger in proportion to the flue gas flow. As a result, in conventional boilers, the amount of steam production possible depends on the feedwater temperature.

Figure 5–28 shows two examples of conventional steam generators with differing feedwater temperatures. It is obvious that with the same difference in temperature at the end of the economizer, the heat available for evaporation and superheating is significantly greater where the feedwater temperature is higher. Therefore, raising the feedwater temperature can increase the amount of live steam produced by a conventional boiler. Further, by preheating the boiler feedwater, the water/steam cycle efficiency is increased because more steam is fully utilized and less energy is lost in the condenser.

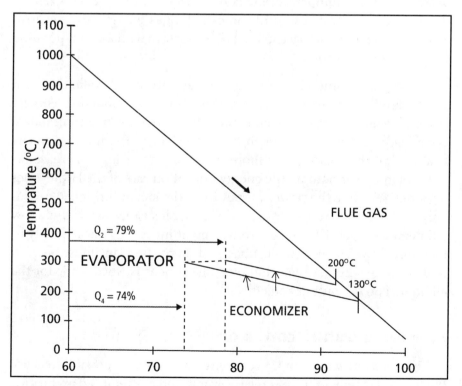

Figure 5–28 Energy/temperature diagram for a conventional boiler

Recirculation

In the single-pressure cycle example, shown in figure 5–18, there is a single stage of preheating in the feedwater tank/deaerator and the feedwater temperature is relatively low at 60°C (140°F). However, the stack temperature is 133°C (271°F) and contains energy that could be used to preheat the feedwater. One way to do this is by increasing the economizer surface and recirculating some of the heated water back to the feedwater tank to preheat the condensate. Upon entering the feedwater tank, the heated water undergoes a sudden pressure drop that causes it to flash partly into steam, which enables both preheating and deaeration to take place. Because it is no longer necessary to extract the steam from the steam turbine for preheating, there is an increase in the steam turbine output. If all of the preheating in the example is done by recirculation, the gross power output of the cycle increases by 630 kW and the gross efficiency from 57.7% to 57.8%. The stack temperature falls from 133 to 112°C (271 to 234°F), demonstrating better usage of the exhaust gas energy. The negative aspect of this is an increase in the HRSG surface with a corresponding increase in HRSG cost.

The larger volume flow of exhaust steam produced could also lead to increased costs in the steam turbine and the condenser/cooling system because their dimensions must be increased to accommodate the additional flow at the steam turbine exhaust. The increase in the heat load of the condenser is more than the proportional increase for the gain in power output. The energy utilization rate of the HRSG rises by about 4% while the power output from the steam turbine increases only by 0.6% because the additional exhaust heat recovered is at a low temperature level. The rate for converting it into mechanical energy is, therefore, low. The economical parameters of the plant must be used to determine whether this additional investment is worthwhile for the additional power output and efficiency.

Low temperature corrosion

This solution with economizer recirculation is only possible because the feedwater temperature is relatively low and the required preheating energy is available in the exhaust gas behind the evaporator. This may not be possible if the feedwater temperature has to be raised to avoid low temperature corrosion in the economizer. This is a chemical limitation on the energetic use of the exhaust gases. The corrosion, caused by water vapor and sulfuric acid in the exhaust gas, occurs whenever the gas is cooled below the acid dew point of these vapors.

In an HRSG, the heat transfer on the exhaust gas side is inferior to that on the steam or waterside. For that reason, the surface temperature of the tubes on the exhaust gas side is approximately the same as the water or steam temperature. If these tubes are to be protected against an attack of low temperature corrosion, feedwater temperature must remain approximately as high as the acid dew point. Thus, a high stack temperature for the exhaust gases does not help if the temperature of the feedwater is too low. Low temperature corrosion can occur even when burning fuels containing no sulfur if the temperature drops below the water dew point. This is generally between 40 and 45°C (104 and 113°F).

The sulfuric acid dew point temperature for a fuel depends on the quantity of sulfur in that fuel. A feedwater temperature of 60°C (140°F), as in figure 5–18, corresponds to a gas fuel with very low sulfur content (<3ppm sulfur in fuel). Oil tends to have more sulfur, resulting in feedwater temperatures of around 120 to 160°C (248 to 320°F). Multistage preheating, as in a conventional cycle, would improve the efficiency, but generally single-pressure cycles are chosen where efficiency is not highly evaluated and further investment would not be economically viable.

For cycles burning fuels with high sulfur content, the easiest improvement to the single-pressure cycle is to use an additional heat exchanger at the end of the HRSG to recover heat for preheating the feedwater. This preheating loop must be designed so that temperatures do not drop below the acid dew point. It is, therefore, normal practice to install an evaporator loop in the HRSG operated at a pressure equivalent to the acid dew point temperature.

Figure 5–29 shows a solution in which a low-pressure evaporator generates saturated steam solely for the feedwater tank/deaerator. In this case, because this loop is at a low pressure at the low temperature end of the HRSG, the power required to drive the additional feedwater pump is quite small.

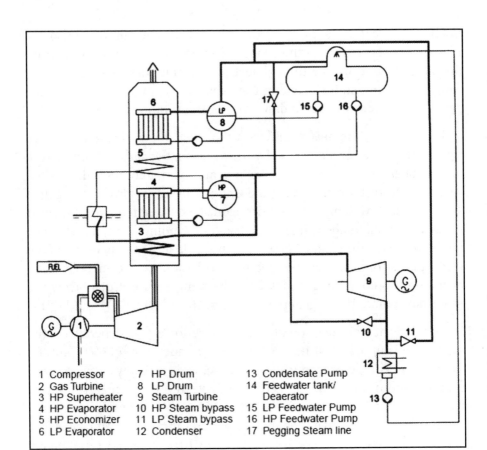

Figure 5–29 Flow diagram of a single-pressure cycle with LP preheating loop for high sulfur fuels

The deaerator/feedwater tank could alternatively be integrated into the drum of the preheating loop, which then functions as a feedwater tank for the cycle in providing a water buffer. This simplifies the cycle mainly in a combined cycle with only one HRSG because additional feedwater pumps and level controls are not required. This kind of configuration is normally used where the fuel has a high sulfur content requiring a high feedwater temperature.

Dual-pressure Cycles

Dual-pressure cycles for high sulfur fuels

A dual-pressure cycle for fuel with high sulfur content is shown in figure 5–30.

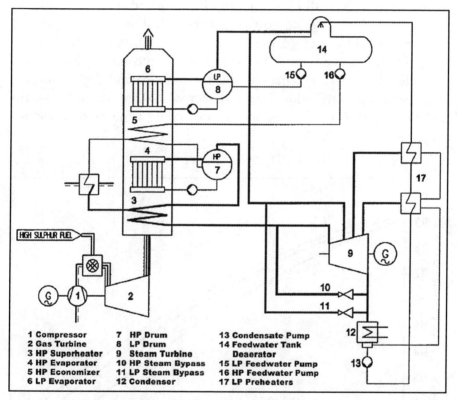

Figure 5–30 Flow diagram of a dual-pressure cycle for high sulfur fuel

The LP evaporator loop (6, 8) generates steam for the steam turbine and for feedwater preheating in the feedwater tank (14). There are two stages of condensate preheating with low-pressure preheaters (17), which are fed by extractions from the steam turbine. This makes sense thermodynamically because the steam used to preheat the feedwater is of low quality. As opposed to the single-pressure cycle with a preheating loop, the excess LP steam is expanded in the steam turbine.

Figure 5–31 shows the effect of the number of preheaters on the steam turbine output for a range of feedwater temperatures. The higher the required feedwater temperature, the lower the steam turbine output in all cases because less energy is available in the HRSG and more steam is required for preheating. Increasing the number of preheating stages increases the steam turbine output because the extracted steam is used more efficiently. To raise the feedwater temperature to a certain level, for example 120°C (248°F), steam must be extracted at a certain level in the steam turbine (e.g., at a pressure corresponding to 130°C (266°F). If this is done in one stage, the entire steam must be extracted at this level. If two stages are used, a second extraction could be done at, for example, 100°C (212°F), enabling this steam to expand slightly further in the steam turbine before being extracted, and so raising the steam turbine output. A similar improvement could be achieved with further stages of preheating.

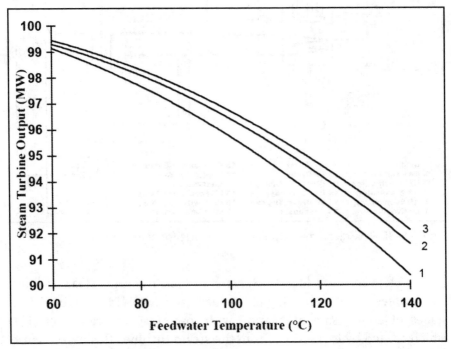

Figure 5–31 Effect of feedwater temperature and number of preheating stages on steam turbine output of a dual-pressure cycle

Dual-pressure cycles for low sulfur fuels

In case of natural gas (fuel with no or low sulfur content) the feedwater temperature can be further reduced. Additional economizers in the HRSG can replace the LP preheaters in figure 5–31. This enables more exhaust gas energy to be utilized by lowering the stack temperature. This is the most common type of dual-pressure cycle (figure 5–32). The first section in the HRSG (7) is a dual-pressure economizer, divided into an LP part for the LP feedwater and an HP part for the first step in heating the HP feedwater.

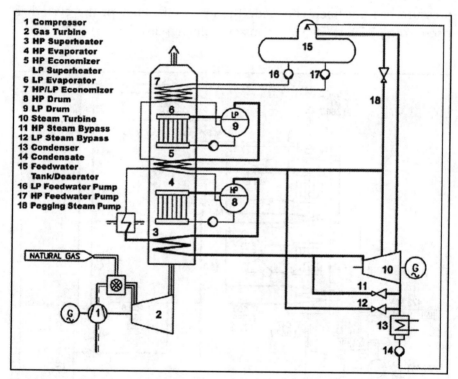

Figure 5–32 Flow diagram of a dual-pressure cycle with low sulfur fuel

More of the steam generated in the LP evaporator and superheater (6 and 5) is available for the dual-admission steam turbine. An extraction is made from the steam turbine for feedwater preheating and deaeration. Pegging steam (18) for off-design conditions for which the steam turbine extraction pressure is too low is taken from the LP live-steam line.

Figure 5–33 shows the heat balance for a dual-pressure cycle burning natural gas with low sulfur content and a feedwater temperature of 60°C (140°F). Comparing it with the single-pressure cycle example, figure 5–18 shows clearly how the dual-pressure cycle makes better use of the exhaust gas in the HRSG, resulting in a higher steam turbine output. The economizers at the end of the HRSG utilize the exhaust gases to a greater extent, lowering the stack temperature to 96°C (205°F) compared to 133°C (272°F) for the single-pressure cycle. The HP part of the HRSG is not affected by the presence of the LP section and HP steam production is the same. Overall steam production is increased due to the 5.4 kg/s, (42,000 lb/hr) of LP steam because the HP steam pressure has not been changed. The gross efficiency of the cycle has risen from 57.7% to 58.6%- a considerable increase.

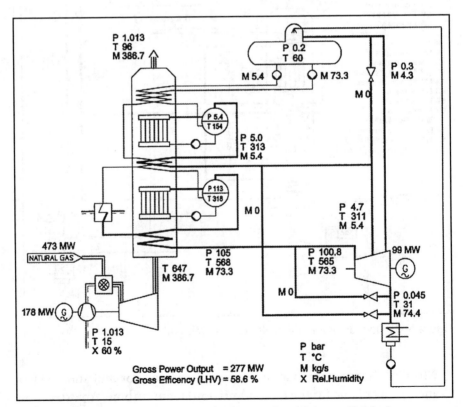

Figure 5–33 Heat balance for a dual-pressure cycle with low sulfur fuel

Figure 5–34 shows the energy flow diagram for the dual-pressure cycle. Compared to the single-pressure cycle (figure 5–19), the stack losses are reduced from 11.4% to 8.2% of the total fuel energy input. However, due to the increased LP steam flow, more energy is lost in the condenser; 32.1% instead of 29.9%. The portion of the steam turbine output has increased by 0.9%. This shows that the additional absorbed heat in the HRSG is converted with a moderate efficiency of 28% because it is being done at a low temperature level.

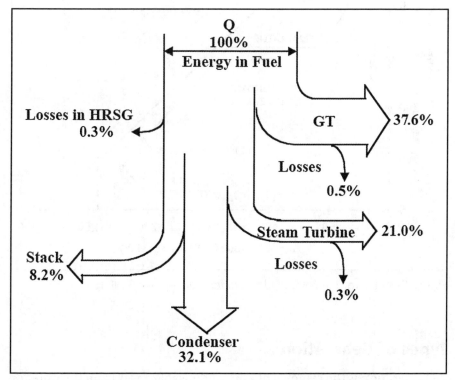

Figure 5–34 Energy flow diagram for a dual-pressure combined cycle plant

Figure 5–35 is the energy/temperature diagram for the dual-pressure HRSG. As it shows, most of the heat exchange takes place in the HP portion of the HRSG. This is directly related to the chosen pressure levels of the cycle. Comparing figure 5–35 with figure 5–17 shows how the energy utilization at the cold end of the HRSG has been improved in the dual-pressure cycle.

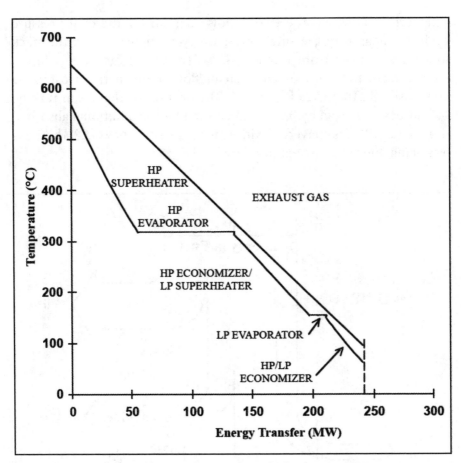

Figure 5–35 Energy/temperature diagram for a dual-pressure HRSG

Types of deaeration

Deaeration is the removal of noncondensable gases from the water or steam in the water/steam cycle. It is very important because high oxygen content can cause erosion/corrosion of the components and piping that come into contact with the water. Typically, an oxygen content of less than 20 parts per billion (ppb) in the feedwater is recommended. Deaeration must be done continuously because small leakages of air at flanges and pump seals in the part of the water/steam cycle (which is under vacuum) cannot be avoided. Indeed, in the design of these components, leakage minimizing is an important factor.

Deaeration takes place when water is sprayed and heated (boiled), thereby releasing the dissolved gases. Generally the condensate is sprayed into the top of the deaerator, which is normally placed on top of the feedwater tank (sometimes on the LP drum). Heating steam, fed into the lower part, rises and heats the water droplets to the saturation/boiling temperature, releasing incondensable gases that are carried to the top of the deaerator and evacuated. The feedwater tank is filled with saturated and deaerated water, and a steam buffer above it prevents any reabsorbtion of air. Depending on the pressure in the feedwater tank/deaerator the deaeration process either takes place under vacuum (known as vacuum deaeration) or at a pressure above atmospheric pressure (known as overpressure deaeration). Both concepts enable good deaeration to take place.

The feedwater system is simpler with vacuum deaeration than with overpressure deaeration because all of the feedwater heating can be done in the feedwater tank by adding the appropriate amount of heating steam without need of additional heat exchangers. Also, the feedwater temperature can be varied for the same system, for example to accommodate oil operation.

With overpressure deaeration, however, as opposed to vacuum deaeration, the noncondensable gases can be exhausted directly into the atmosphere, independently of the condenser evacuation system.

Both of the previously mentioned examples (single and dual pressure) have been done with vacuum deaeration. The feedwater temperature is at 60°C (140°F), corresponding with a pressure in the feedwater tank/deaerator of 0.2 bar (2.9 psia). In contrast, the example with the preheaters has overpressure deaeration at 135°C, 3.1 bar (275°F, 45 psig).

Deaeration will also partly take place in the condenser. The process is similar to that in the deaerator. The steam turbine exhaust steam condenses and collects in the condenser hotwell, while the noncondensable gases are extracted by means of evacuation equipment. Again, a steam cushion separates the water in the hotwell so that reabsorbtion of the air cannot take place. Often, levels of deaeration in the condenser can be achieved that are comparable to those in the deaerator. Therefore, the dedicated deaerator/feedwater tank can sometimes be eliminated from the cycle, and the condensate is fed directly from the condenser into the HRSG drum. In these cases the makeup water must be admitted to the cycle through the condenser. For cycles of this type, the condenser hotwell capacity must be increased (or the HRSG LP drum) to provide the cycle with a water buffer because there is no feedwater tank. To raise the condensate temperature above the sulfur dew point temperature at the HRSG inlet, feedwater is recirculated from the HRSG to a point upstream of the economizer inlet.

An important factor in determining which type of deaeration to use is the quantity of makeup water and where it is to be admitted to the cycle. Makeup water is usually fully saturated with oxygen, but if the required continuous quantity is less than about a few percent of the condensate flow, condenser deaeration may be appropriate with the makeup water being sprayed into the cycle in the condenser neck (above the tube bundles). For cycles with process extractions a separate (makeup water) deaerator is normally preferred.

Main Design Parameters of the Dual-Pressure Cycle

Live-steam pressure

Two items must be noted when selecting the HP and LP live-steam pressures for dual-pressure cycles. The HP steam pressure must be relatively high to attain good exergetic utilization of the exhaust gas heat; and the LP steam pressure must be low to attain good energetic and exergetic utilization of the exhaust gas heat and, hence, to achieve a higher steam turbine output.

Figure 5–36 shows the steam turbine output for the dual-pressure cycle as a function of the HP steam pressure for various LP steam pressures. Other parameters remain constant. As either pressure rises, the steam turbine power output increases for a given steam mass flow because of the increased enthalpy drop in the steam turbine. However, at higher pressures, less steam is generated so there is a trade off between the higher enthalpy drop and lower mass flow. This is more pronounced for the LP section due to the relatively big influence that a change in pressure will have.

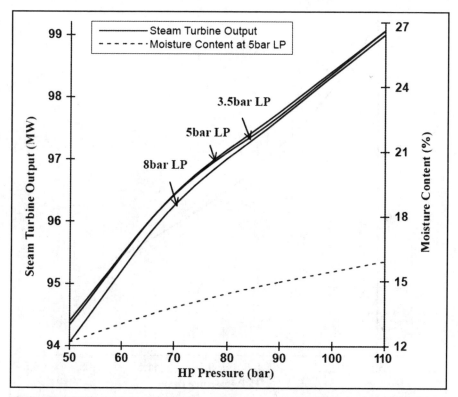

Figure 5–36 Effect of the HP and LP pressure on steam turbine output and exhaust moisture content for a dual-pressure cycle

The pressure in the LP evaporator should not be below about 3 bar (29 psig) because the enthalpy drop available in the turbine becomes very small, the volume flow of steam becomes very large, and the hardware is, therefore, more expensive. LP pressure is chosen at 5 bar (58 psig). The moisture content also rises with the HP pressure. This is why the HP pressure is not increased beyond 105 bar (1508 psig) for an LP pressure of 5 bar (58 psig) as in figure 5–33.

Figure 5–37 shows how the HRSG efficiency improves as the LP pressure decreases. However, for thermodynamic reasons, the best energy utilization of the HRSG does not necessarily result in the highest steam turbine output as was seen for the single-pressure cycle.

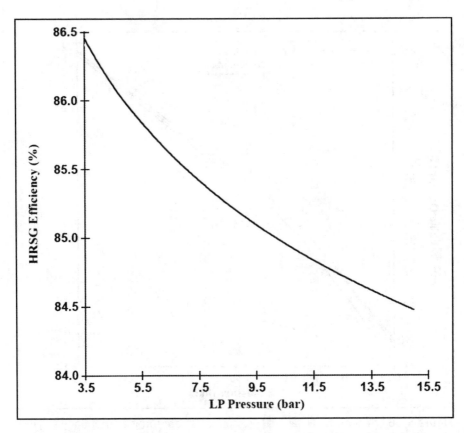

Figure 5–37 Effect of LP pressure on HRSG efficiency for a dual-pressure cycle

Live-steam temperatures

Unlike the single-pressure cycle, in the dual-pressure cycle a live-steam temperature increase brings with it a substantial improvement in the output. The same arguments as for the single-pressure cycle are valid for the HP section, where a slight decrease in output for an increase in temperature was seen. However, in the dual-pressure cycle, more energy is made available to the LP section if the HP temperature is raised, which more than compensates for the slight output loss of the HP steam.

In figure 5–38 the extent of this improvement on steam turbine output is shown. In the LP section, as opposed to the HP section, a high degree of superheat improves the output marginally. The curve was calculated with a constant pressure drop in the LP superheater. In reality, this pressure drop would be in proportion to the degree of superheat raising the curve slightly towards point A. A cycle without a superheater is advantageous for the simplicity of the system. The benefit of increasing the HP temperature is significant for the steam turbine output. Increasing the LP temperature is important because it leads to a reduction in the moisture content in the last stage of the steam turbine. There is also a reduction in the HRSG surface because although the LP superheater surface increases, those of the LP evaporator and economizer will decrease due to the smaller mass flow, thus the net effect is a decrease.

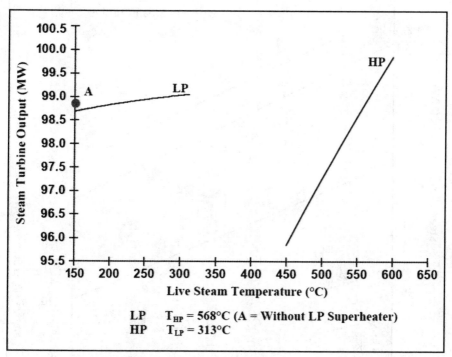

Figure 5–38 Effect of HP and LP steam temperature on steam turbine output for a dual-pressure cycle

Pinch point

The pinch point of the HP evaporator has less influence here than in a single-pressure cycle because energy that was utilized in the HP section can be recovered in the LP section of the HRSG. Transferring energy from the HP to the LP section by increasing the HP pinch point causes exergetic losses. This is because HP steam is far more valuable than LP steam due to the greater enthalpy drop in the steam turbine. Figure 5–39 shows the steam turbine output and relative HRSG surface as a function of the pinch points of the HP and LP evaporators.

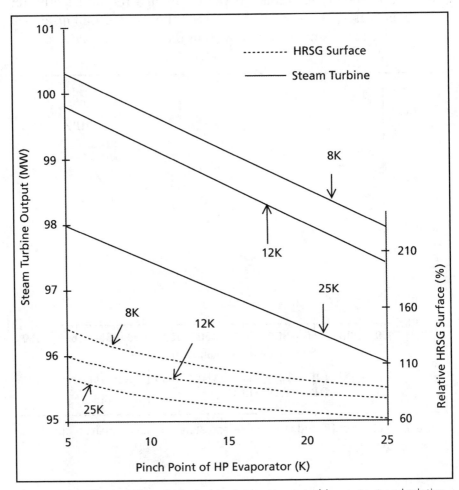

Figure 5–39 Effect of HP and LP pinch point on steam turbine output and relative HRSG surface for a dual-pressure cycle

The HP and LP pinch points are interrelated. A reduction of the HP pinch point increases the surface of the HP evaporator and economizer, but also reduces the one of the LP section. This is because the exhaust gas temperature after the HP economizer falls, reducing the amount of heat available for the LP section and, therefore, the LP steam flow. The heat required for LP feedwater heating decreases because the LP feedwater flow is smaller.

For a dual-pressure cycle, the pinch points of both the HP and the LP evaporators have less effect on the output of the steam turbine than the pinch point in a single-pressure cycle. If equal economic value is attached to the efficiency, the pinch points selected for dual-pressure cycles should be larger than those for single-pressure cycles. However, dual-pressure cycles are selected only where efficiency is valued more highly, which means that the pinch points should remain similar.

Triple-pressure cycle

If a third pressure level is added to the dual-pressure cycle, a further improvement can be made, mainly by recovering more exergy from the exhaust gas. The flow diagram is shown in figure 5–40. Separate pumps (20 and 21) supply feedwater to a dual-pressure economizer (9), one at the HP, the other one at the intermediate pressure (IP) level. On leaving the economizer, the IP feedwater divides into two parts, one entering a second dual-pressure economiser (7) and the other one being throttled into the LP steam drum (12). The saturated LP steam collecting in the drum is fed directly to the steam turbine.

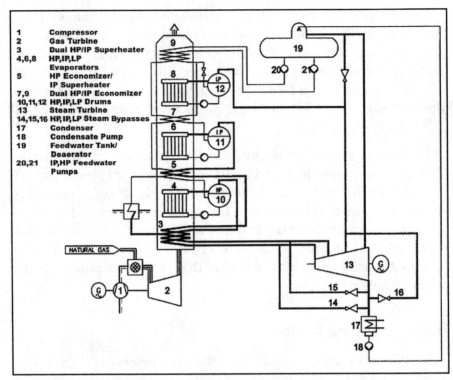

Figure 5–40 Flow diagram of a triple-pressure cycle

The HP and IP pressure levels follow the pattern of economizer, evaporator, and superheater until the superheated steam generated is fed to the triple-pressure steam turbine. Again each pressure level has a separate steam turbine bypass (14, 15, 16).

Figure 5–41 shows the heat balance for a triple-pressure cycle with a low sulfur natural gas fuel and a feedwater temperature of 60°C (140°F).

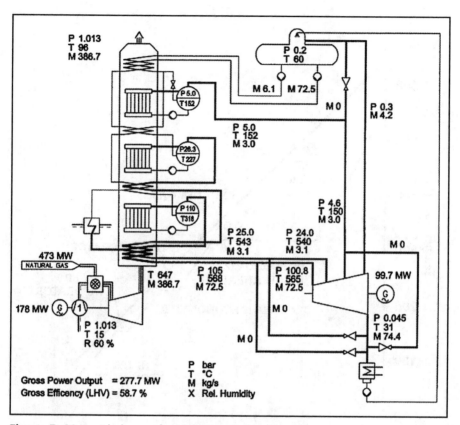

Figure 5–41 Heat balance of a triple-pressure cycle

For this example there is only a slight improvement in output over the dual-pressure cycle, improving efficiency marginally from 58.6% to 58.7%. The stack temperature is the same. The improvement is due to a slight exergetic gain caused by the IP level, reducing the area between the exhaust gas and the water/steam lines on the temperature/energy diagram figure 5–42.

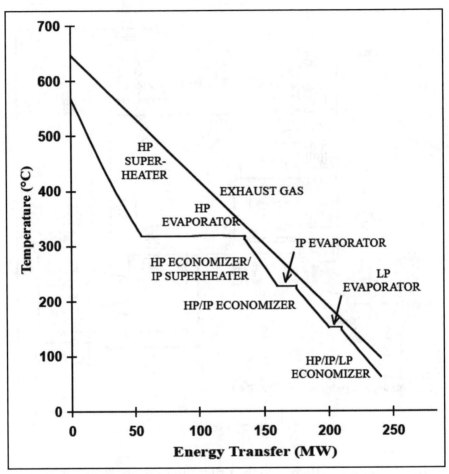

Figure 5–42 Energy/temperature diagram of a triple-pressure HRSG

Compared to the HP flow 72.5 kg/s (575,200 lb/hr), the IP and LP flows are very small, only 3.1 kg/s (24,600 lb/hr) and 3.0 kg/s (23,800 lb/hr), respectively. This limits the potential gain in exergy due to the addition of the IP section. The HP flow is slightly less (at constant live-steam pressure) than the one for the dual-pressure example because the IP superheater is at a higher temperature level, removing energy from the HP section of the HRSG. It results in a minor negative effect on output, and is part of the reason why the benefit in this triple cycle is so small. A benefit is gained because IP steam is generated in the place of some LP steam.

The energy flow diagram, shown in figure 5–43, shows the slight improvement in the steam turbine output over the dual-pressure cycle for the reasons previously given.

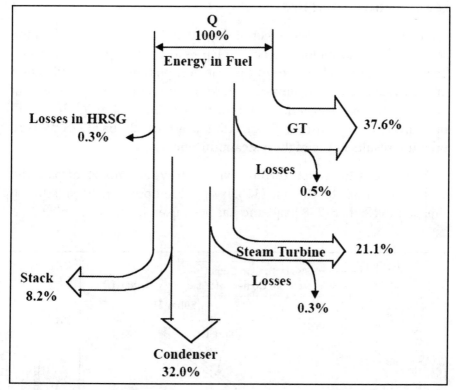

Figure 5–43 Energy flow diagram of a triple-pressure combined-cycle plant

The stack losses (stack temperature) are the same; however, there is a slight decrease in heat load in the condenser that corresponds to the gain in steam turbine output. For gas turbines with lower exhaust gas temperatures, more energy would be available for IP steam production because the HP steam production would be lower, thus making this concept more attractive.

Main Design Parameters of the Triple-Pressure Cycle

Live-steam pressure

Figure 5–44 shows the effect of the HP and IP pressure level on the steam turbine output for a constant LP pressure. These pressures should be optimized to such an extent that the best possible exergy utilization of the exhaust gas is achieved. At low HP pressures an IP pressure of 10 bar (130 psig) brings more output than 30 bar (420 psig). If the HP pressure is increased to 105 bar (1508 psig), the 20 bar (275 psig) IP pressure results in the highest steam turbine output.

Moisture at the steam turbine exhaust plays an important part and sets the HP limit to 105 bar (1508 psig). The optimum IP pressure at this point is 25 bar (248 psig) and has been used for the example.

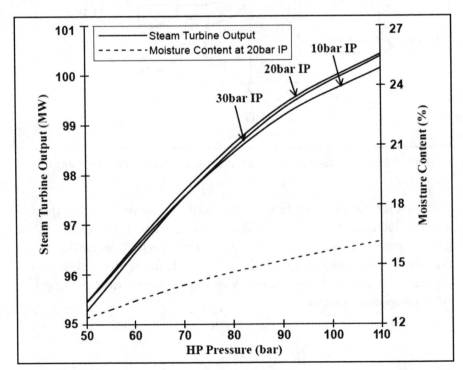

Figure 5–44 Steam turbine output and exhaust moisture content versus HP and IP pressure for triple-pressure cycles at constant LP pressure (5 bar)

For a given HP and IP pressure, the maximum steam turbine output is at a clearly definable LP pressure, as can be seen from figure 5–45. For the dual-pressure cycle, however, LP pressures below 3 bar (29 psig) are not recommended. The HRSG surface decreases as the LP pressure increases because less heat exchange takes place at the low temperature end of the HRSG, thus reducing the HRSG cost. In the example (Fig 5–41), the LP pressure was chosen to be 5 bar (58 psig).

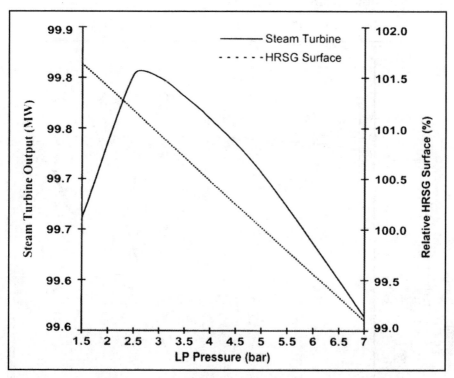

Figure 5–45 Effect of LP pressure on steam turbine output and relative HRSG surface for triple-pressure cycles at constant HP (105 bar) and IP (25 bar)

Live-steam temperature

Figure 5–46 shows the relationship between live-steam temperatures and steam turbine output. The HP steam temperature has a significant effect, whereas higher IP and LP steam temperatures only slightly increase the power output. There is, however, another advantage with a high IP temperature:

By heating the IP steam to a temperature level close to the one of the HP live steam, a small reheat effect is seen on the steam turbine expansion line. As a result, the hot IP steam mixes with the HP steam that has already expanded in the HP turbine. It so decreases the moisture content and, therefore, the risk of erosion due to high wetness in the last stage of the steam turbine.

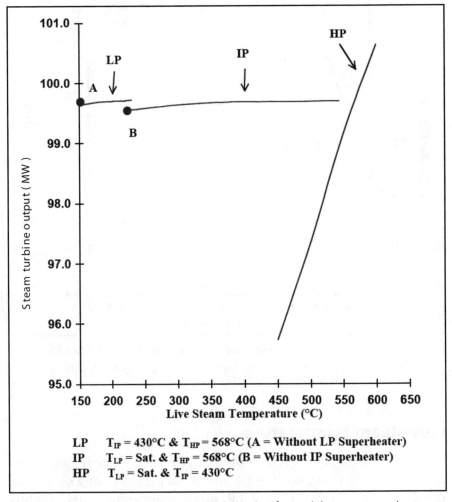

Figure 5–46 Live-steam temperature optimization for a triple-pressure cycle

Figure 5–47 shows this effect on an enthalpy/entropy diagram for steam. The mild reheat causes the steam turbine expansion line to move to the right where wetness is lower. However, there is no improvement in moisture content in this triple cycle compared with the dual-pressure cycle because any benefit derived from the IP superheater is eliminated by the removal of the LP superheater.

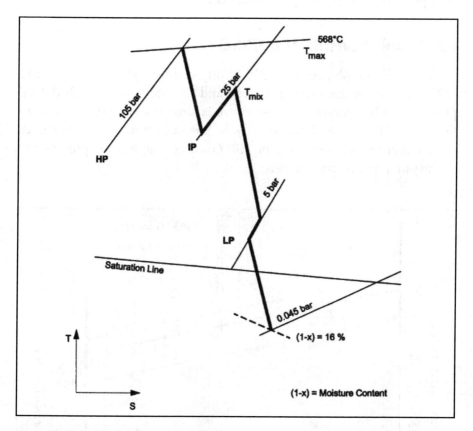

Figure 5–47 Temperature/entropy diagram showing the effect of mild reheat on the steam turbine expansion line

Care must be taken to limit the thermal stresses within the steam turbine, arising from a temperature difference between the HP steam after expansion and the IP live steam at the mixing point. It can be solved by the constructional design of the steam turbine casing at the IP steam admission point.

For a modern gas turbine, similar to the one used in the example, the potential gain due to this mild reheat effect is limited because of the low quantity of IP steam (approximately 5% of the HP steam flow). It also limits the temperature increase at the mixing point in the steam turbine. A similar mild reheat effect would be gained if there were a superheater in the LP section. However, this would make the cycle more complex with practically no gain in performance.

Pinch point within the HRSG

Figure 5–48 shows the effect on steam turbine output and HRSG surface when varying the HP and IP pinch points with the LP pinch point held constant. This curve is similar to the one shown in figure 5–39 for the dual-pressure cycle. Steam turbine output increases with reduced pinch point and the HRSG surface increases exponentially as the pinch points tend to zero.

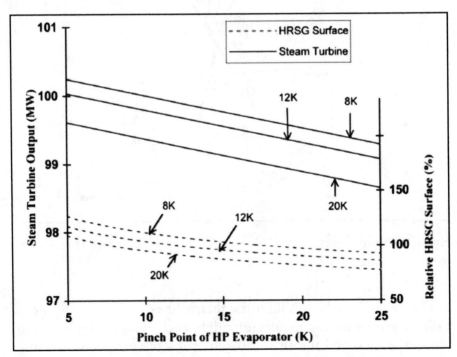

Figure 5–48 Effect of HP and IP evaporator pinch point on steam turbine output and relative HRSG surface for a triple-pressure cycle with constant LP pinch point

Reheat Cycles

From our analysis thus far, it is clear that the moisture content in the steam turbine exhaust is the significant factor in limiting further improvements in the performance of the various cycles. Extending the idea of mild reheat results in a full reheat cycle, with the advantage of a reduced moisture content and an improvement in the performance. Taking the triple-pressure cycle as example, after the expansion of the HP steam in the steam turbine to IP level, the steam returns to the HRSG and mixes with the steam leaving the IP superheater. This steam is then heated to a temperature similar or equal to the HP live-steam temperature, before admitting it to the steam turbine. In a dual-pressure reheat cycle, there is no mixing on re-entering the HRSG; instead, the cold reheat goes to an independent reheater section.

The reheater takes more exergy out of the HRSG and increases the steam turbine enthalpy drop, resulting in a higher steam turbine output. Because the last part of the steam turbine expansion line is moved further to the right on the enthalpy/entropy diagram, there is a reduction in the moisture content.

Triple-pressure reheat cycle

Figure 5–49 shows the flow diagram of the triple-pressure cycle with reheat. The steam turbine has separate HP and IP/LP casings (13 and 14), to accommodate the extraction of the cold reheat. The high-pressure bypass (15), instead of dumping steam into the condenser, dumps it into the cold reheat line, such as the line leaving the HP steam turbine.

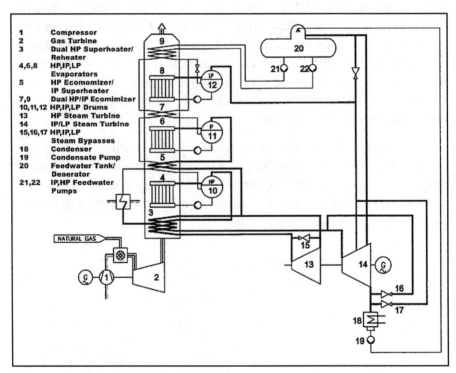

Figure 5–49 Flow diagram of a triple-pressure reheat cycle

The heat balance is shown in figure 5–50. The HP live-steam pressure now increased to 120 bar (1,725 psig), together with the reheater (which removes energy at the hot end of the HRSG) leads to a reduced HP mass flow of 59.2 kg/s (469,700 lb/hr) compared to the triple-pressure cycle. The IP steam flow is, for that reason, slightly higher 5.9 kg/s (46,800 lb/hr). However, there is a significant improvement in the cycle performance because, due to the reheat, there is a greater exergy transfer in the hot end of the HRSG. The IP steam (IP and reheat steam) is expanded from a high temperature level of 65.1 kg/s (516,500 lb/hr) at 565°C (1,049°F) instead of 75.6 kg/s (599,800 lb/hr) HP and IP steam) with a mix temperature of 354°C (669°F) in the triple-pressure cycle. The resulting gross output is 2.8 MW higher than for the triple-pressure cycle with an efficiency of 59.3% instead of 58.7%.

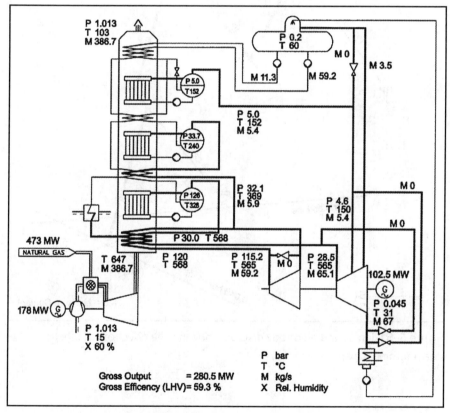

Figure 5–50 Heat balance for a triple-pressure reheat cycle

The decrease in the moisture content is illustrated on the enthalpy/entropy diagram of figure 5–51. The value is 10% compared to 16% in the triple-pressure example. The percentage of the stack losses has increased by 0.4% (figure 5–52) because the stack temperature is slightly higher 103°C (217°F).

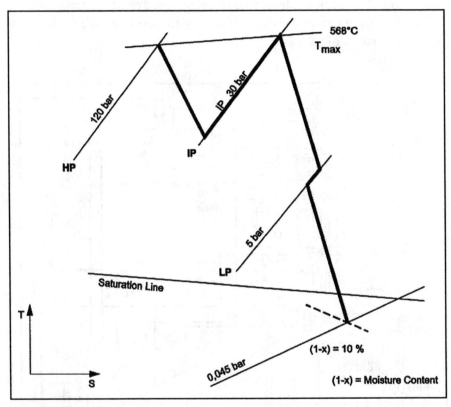

Figure 5–51 Temperature/entropy diagram showing the effect of full reheat on the steam turbine expansion line

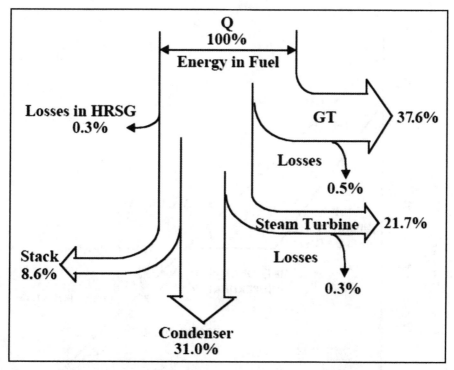

Figure 5–52 Energy flow diagram for a triple-pressure reheat combined-cycle plant

This occurs because the reheater steals energy from the HP section of the HRSG, resulting in less HP steam production and, therefore, less feedwater heating in the HP economizers. Increases in IP and LP mass flows do not compensate for this, and so the stack temperature increases. Combined with the additional steam turbine output, it explains the lower heat load in the condenser.

The energy/temperature diagram for the triple-pressure reheat cycle is shown in figure 5–53. The energy taken out in the HP superheater/reheater and HP evaporator is approximately 138 MW compared to 140 MW for the triple-pressure cycle. This is mainly due to the HP evaporator pressure level difference. However, the energy is transferred at a higher temperature level and the increased combined mass flow of the HP and reheater HRSG sections (compared to the one of the triple-pressure cycle) move this part of the diagram closer to the exhaust gas temperature. It is illustrated by the area between exhaust gas and water/steam lines, results in an exergy gain and, therefore, a higher steam turbine output.

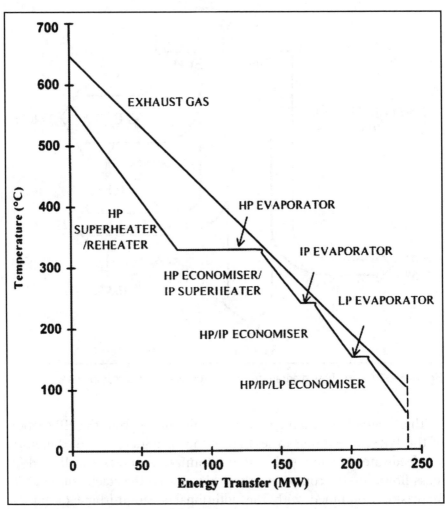

Figure 5–53 Energy/temperature diagram for a triple-pressure reheat HRSG

Live-steam data

The effect of live steam and reheat pressure on the steam turbine output is shown in figure 5–54. Power increases with live-steam pressure, although the improvement is less after a certain point, depending on the reheat temperature. The IP pressure is chosen to match the reheat pressure after accounting for losses in the reheater and cold reheat piping.

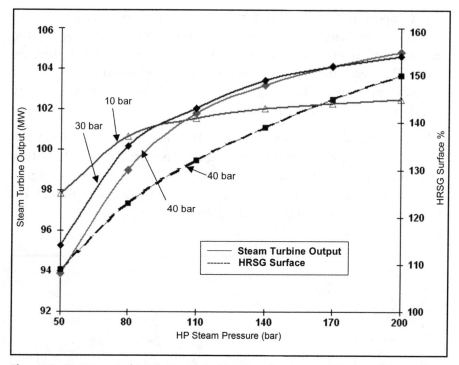

Figure 5–54 Steam turbine output and HRSG surface versus HP and reheat pressure for a triple-pressure reheat cycle at constant HP and reheat temperature (568/568°C)

An increase in the HP pressure leads to an increase in the cost of the HRSG, feedwater pumps and piping, which is often the limiting factor when determining the correct pressure. There will also be an increase in the auxiliary consumption of the feedwater pumps as the HP pressure rises, which means that the benefit in the net power output is slightly less than shown on the curve in figure 5–54.

At low live-steam pressures a lower reheat pressure is advantageous because the enthalpy drop in the HP steam turbine is bigger. The two pressures should not be too close together, as illustrated by the point with 50 bar live-steam pressure and 40 bar reheat pressure. Similar to the HP pressure increase, a higher reheat pressure is preferable. As shown in figure 5–54 the optimal reheat pressure for a HP steam pressure of 80 bar is around 10 bar, for a HP steam pressure of 170 bar it will be around 40 bar.

It is important to keep the steam-side pressure loss in the cold reheat piping, reheater, and hot reheat piping as low as possible. Increasing the pressure loss has the same effect as throttling over the steam turbine valves. The steam turbine expansion line is moved towards the right on the enthalpy/entropy diagram, resulting in a higher exhaust enthalpy and, therefore, a lower steam turbine output.

Figure 5–55 shows the effect of the HP live-steam and reheat steam temperature on the steam turbine output. An increase in temperature increases the steam turbine output significantly. The curve in figure 5–55 shows both temperatures being varied at the same time, each contributing approximately half the gain.

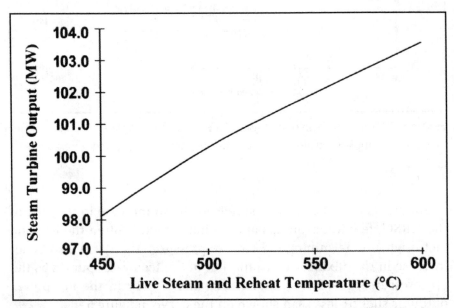

Figure 5–55 Steam turbine output versus HP and reheat steam temperature for a triple-pressure reheat cycle at constant HP (120 bar), IP (30 bar) and LP (5 bar) pressure

The relationship between the pinch points of these cycles is the same as the one for the triple-pressure non-reheat cycle.

High-pressure reheat cycles (dual-pressure reheat)

It is clear that there is a performance benefit in a reheat cycle because of the larger expansion in the steam turbine and the improved exergy utilization in the HRSG. Still more can be achieved if the live-steam pressure can be raised. It has been shown that for dual- and triple-pressure cycles, the HP part of the cycle makes the main contribution to the performance; therefore, the major investment should be placed here.

Figure 5–56 shows the flow diagram of a cycle, which has been derived with these factors in mind, to achieve a high efficiency at a lower level of plant complexity. It is a high-pressure reheat cycle and a dual-pressure cycle because the aim was to improve the exergy exchange in the HP part of the HRSG. To accommodate the higher pressure, a once-through HRSG HP section (2 and 4) is chosen. Unlike the conventional drum system, there is no drum, and the economizer, evaporator, and superheater form an integral HRSG part. The separator (5) ensures safe operation and control of the HRSG. It will, for example, prevent water from entering the superheater, and provides a means of blow down for the HP section during low loads and transient conditions. The separator is dry in normal operating conditions. The same equipment is used in conventional plants with once-through boilers. The LP part of the HRSG is of the drum-type design. The increased HP pressure improves the exergy transfer in the HP section, although this improvement is partly reduced because there is no IP section, which is omitted to reduce the complexity of the cycle. For steam turbine bypass operation, the reheater is run dry and all HP steam is dumped into the condenser.

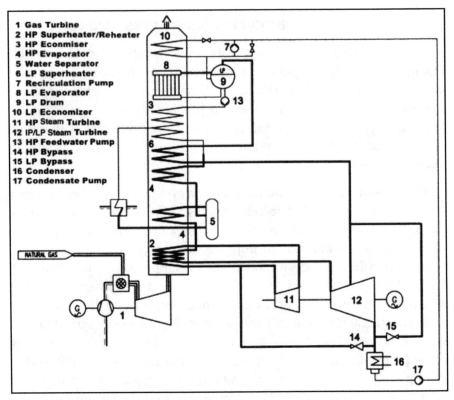

Figure 5–56 Flow diagram of a high-pressure reheat cycle with a HP once-through HRSG and a drum-type LP section

This cycle has no feedwater tank, so deaeration takes place in the condenser. A recirculation loop in the LP economizer of the HRSG is used to raise the temperature of the incoming feedwater to a minimum required temperature level. The condensate pumps (17) must be designed to deliver the condensate to the LP drum, fulfilling the function of LP feedwater pumps as well. The HP feedwater pumps (13) take suction from the LP drum.

Such a concept with an optimized dual-pressure reheat for gas turbines with high exhaust gas temperatures can achieve with a quite simpler cycle (less components, piping, etc.) almost the same efficiency level as traditional triple-pressure reheat cycles. The example is with a fuel that has no sulfur content, so the feedwater temperature has been reduced to optimize the energy in the cold end of the HRSG. In such a cycle the feedwater temperature has an effect on the HRSG surface, but not on the performance, as long as there is sufficient energy available in the exhaust gas to preheat the incoming condensate.

Cycles with Supplementary Firing

Supplementary firing is a way of increasing the plant output by installing duct burners in the HRSG. They add energy to the cycle by increasing the exhaust gas temperature, often at the expense of efficiency.

Supplementary firing is appropriate for HRSG's because there is usually sufficient oxygen content in the exhaust gas to act as combustion air. In an open-cycle gas turbine with a single stage of combustion, only 30 to 50% of the oxygen contained in the air is used for combustion.

Earlier combined-cycle installations generally had supplementary firing. This is not the case today because of progress in the development of the gas turbine. As gas turbine inlet temperatures and, hence, the exhaust temperatures increase, the importance of supplementary firing diminishes for two reasons. The temperature window between the gas turbine exhaust and the duct burner exhaust temperature decreases, so the added benefit of supplementary firing decreases, too. Also, optimum values can be given to the water/steam cycle parameters with the gas turbine alone. Because current levels of efficiency are set by reheat cycles, the efficiency achieved in cycles with supplementary firing lead to lower efficiencies and, therefore, higher cost of electricity.

Nevertheless, increased operating and fuel flexibility of the combined cycle with supplementary firing may be an advantage in special cases, particularly in installations used for cogeneration of heat and power where this arrangement makes it possible to control the electrical and thermal outputs independently (see chapter 6).

Figure 5–57 shows the energy/temperature diagrams for a single-pressure HRSG with constant live-steam conditions and inlet exhaust gas temperatures of 647°C (1,197°F), 750°C (1,382°F) and 1,000°C (1,832°F), the latter two after supplementary firing. At 647°C the temperatures of gas and water in the economizer are convergent, with the minimum difference in temperature on the evaporator end (typical for a HRSG without supplementary firing). At 1,000°C (1,832°F), on the other hand, the minimum difference in temperature is at the inlet to the economizer on the waterside. This pattern corresponds more to the one of a conventional steam generator.

A temperature of 750°C (1,382°F) after supplementary firing provides the best exergetic utilization of the exhaust gas with a constant difference in temperature along the entire economizer. The single-pressure cycle is here at an optimum with the exhaust cooled down to a temperature close to feedwater temperature. There is no exergy or energy available for additional pressure levels.

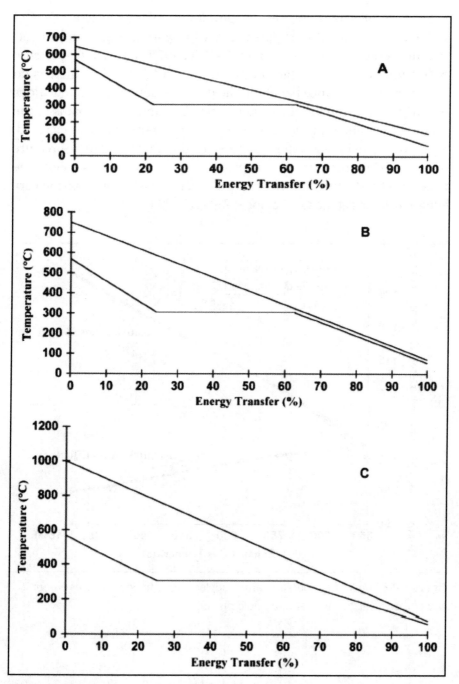

Figure 5–57 Energy/temperature diagram for 647°C (A), 750°C (B) and 1,000°C (C) exhaust gas temperature entering the HRSG

Figure 5–58 shows how relative power output and efficiency depend on the temperature after supplementary firing for a single- and a dual-pressure cycle. The reference point at 647°C (1,197°F) is the single-pressure cycle without supplementary firing. Above 750°C (1,382°F) there is no longer any performance benefit in the dual-pressure cycle. Calculations assume the use of natural gas fuel. When burning oil, the paths of the curves for the single-pressure system would not be significantly changed, but there would be less difference between the single- and dual-pressure systems. Triple pressure and triple-pressure reheat cycles follow the same pattern, starting out with a larger difference without supplementary firing and ending at the same point at 750°C (1,382°F).

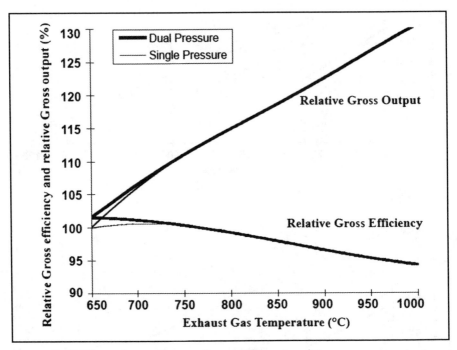

Figure 5–58 Effect of temperature after supplementary firing on power output and efficiency relative to that of a single-pressure cycle

Chapter 5 Combined-Cycle Concepts • 129

For installations with supplementary firing that are frequently operated at part loads, it could make economic sense to select a cycle based on a design point without supplementary firing, consequently with more pressure levels or even reheat. The load point with supplementary firing on is, in such a case, an off-design point. At part load or when the supplementary firing is switched off, the stack temperature rises and the LP section would enable this exhaust gas energy to be used.

Figure 5–59 shows the heat balance of a typical combined-cycle plant with supplementary firing to 750°C (1,382°F) with natural gas and a feedwater temperature of 60°C (140°F). The basic arrangement for this installation is the same as the one for the single-pressure system in figure 5–18.

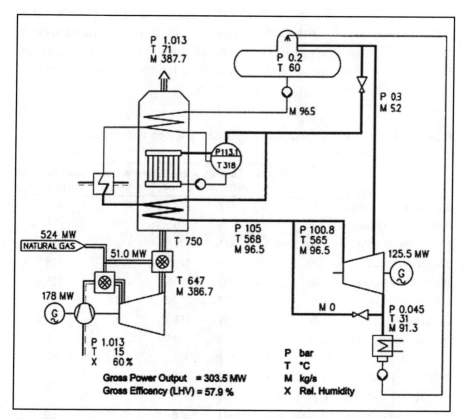

Figure 5–59 Heat balance for a single-pressure cycle with supplementary firing

Supplementary firing increases the steam turbine output by 24.3 MW compared to the cycle without supplementary firing. The efficiency rises slightly from 57.7 to 57.9% because the increased steam production also results in increased mass flows through the economizers removing more energy from the exhaust gas, thus lowering the stack temperature. Generally, for cycles with more pressure levels, supplementary firing has a negative effect on the efficiency because without supplementary firing these cycles already make maximum use of the exhaust gas energy.

Summary of Cycle Performance

To give an overview of all the cycles discussed in this section, the performance data for the examples in this section are summarized in table 5–4. The auxiliary power consumption for each case is also given, as well as the resulting net output and net efficiency, which are crucial for comparison of the cycles in real terms.

Table 5–4 Performance comparison of different cycle concepts (natural gas with low sulfur content and GT exhaust gas temperature of 647 °C (1,197 °F)

		Single-pressure	Dual-pressure	Triple-pressure	Triple-pressure reheat	Single-pressure/suppl. Firing
Gas Turbine Fuel Input (LHV)	MW	473	473	473	473	473
Duct Burner Fuel Input (LHV)	MW	0	0	0	0	51
Total Fuel Input (LHV)	MW	473	473	473	473	524
Gas Turbine Output	MW	178	178	178	178	178
Steam Turbine Output	MW	94.8	99	99.7	102.5	125.5
Gross Plant Output	MW	272.8	277	277.7	280.5	303.5
Gross Efficiency (LHV)	%	57.7	58.6	58.7	59.3	57.9
Auxiliary Consumption	MW	4.1	4.5	4.5	4.6	5
Net Plant Output	MW	268.7	272.5	273.2	275.9	298.5
Net Efficiency (LHV)	%	56.8	57.6	57.8	58.3	57
Net Heat Rate (LHV)	kj/kWh	6337	6249	6233	6712	6320
Net Heat Rate (LHV)	Btu/kWh	6006	5923	5908	5850	5990
Total Relative Plant Cost	%	100	104	106	112	

As illustrated in table 5–4 and explained earlier in this chapter, the performance difference from a dual-pressure cycle to a triple-pressure cycle is quite low. It is, therefore, obvious that a further increase from a triple cycle to a four-pressure-level cycle makes no sense from an economical point of view.

Water/Steam Cycle Evaluation

When defining a water/steam cycle concept for a power plant, there is always a trade off among performance, required investment, and economic criteria. Performance and investment are directly related to each other; as the complexity of a cycle increases, there is generally an increase in performance, but also in the required investment. The exact relationship depends on the type of gas turbine, or rather, the exhaust gas conditions of the gas turbine entering the water/steam cycle.

Figure 5–60 shows the relative performance (efficiency and power output) for different concepts plotted against the GT exhaust gas temperatures.

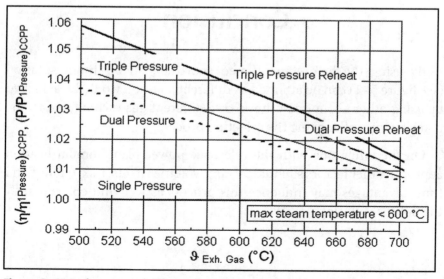

Figure 5–60 Performance of different combined cycles over the exhaust gas temperature

The trends of the curves are similar, however, with higher GT exhaust gas temperatures the benefits from additional pressure levels are getting smaller, for example additional pressure levels to GTs with high exhaust gas temperature can improve the efficiency by almost 4%, whereas the improvement for the turbine with lower exhaust gas temperature of 525°C (977°F) is almost 6%. This is due to the second type of gas turbine; there is more low-grade heat available for additional pressure levels in the HRSG.

Any of the concepts could be appropriate for a power plant, and the optimized choice depends on the economic criteria for the specific project conditions. These are mainly the criteria related to the cost of electricity described in chapter 3. The example using equation (3–8) shows that for an assumed set of economic criteria, it is worthwhile investing up to 4.4% more capital to achieve a 1% improvement in efficiency and output. Applying these criteria to the gas turbine in the examples given later, figure 5–60 shows that this would lead to a triple-pressure reheat cycle.

A very rough cost comparison for the different cycle concepts is given in table 5–4.

Conclusion

By using the before mentioned approach, an optimum solution (see figure 5–1) can be attained. A different solution would be achieved by using other economic criteria or a different gas turbine due to the many factors influencing the selection process.

On the other side with the reference power plant/standardization approach other factors such as learning/volume effects, risk, quality, and time advantages may influence this process as well and consequently the optimal solution.

Future Trends

Together with the ongoing development of the gas turbines, changes in fuel gas prices, and market conditions, the concepts of combined cycles with their design parameters (steam temperature, steam pressure, pinch points, etc.) can be further optimized. Figure 5–61 shows how a change in different parameters/measures can influence the overall output and efficiency of a combined-cycle power plant.

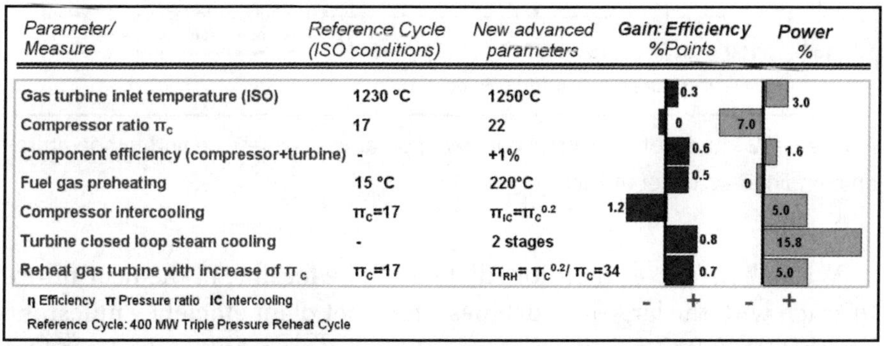

Parameter/ Measure	Reference Cycle (ISO conditions)	New advanced parameters	Gain: Efficiency %Points	Power %
Gas turbine inlet temperature (ISO)	1230 °C	1250°C	0.3	3.0
Compressor ratio π_c	17	22	0	7.0
Component efficiency (compressor+turbine)	-	+1%	0.6	1.6
Fuel gas preheating	15 °C	220°C	0.5	0
Compressor intercooling	$\pi_c=17$	$\pi_{IC}=\pi_c^{0.2}$	1.2	5.0
Turbine closed loop steam cooling	-	2 stages	0.8	15.8
Reheat gas turbine with increase of π_c	$\pi_c=17$	$\pi_{RH}=\pi_c^{0.2}/\pi_c=34$	0.7	5.0

η Efficiency π Pressure ratio IC Intercooling
Reference Cycle: 400 MW Triple Pressure Reheat Cycle

Figure 5–61 Influence of various parameters/measure on combined-cycle output and efficiency

As explained and shown in this chapter, the steam temperature and steam pressure of the HP steam has a big impact on the overall plant efficiency. With new larger gas turbines and an optimized steam turbine concept, a further increase in the live-steam parameter can bring another gain in overall net plant efficiency, as illustrated in figure 5–62.

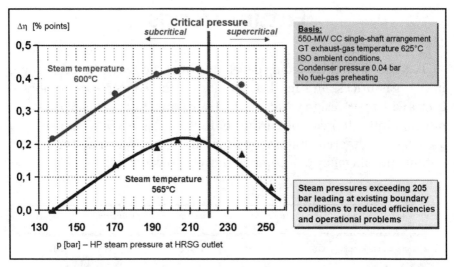

Figure 5–62 Impact of HP and reheat steam parameters (pressure and temperature) on combined-cycle net efficiency

With all measures mentioned, it would be feasible in the near future to reach with the large gas turbines a new net plant efficiency milestone level of 60% at ISO conditions. Nevertheless, it must be clearly stated that the above-mentioned efficiency level does not apply to all plants. Only a clear economical evaluation and analysis of all customer and site related requirements as described will lead to an optimal plant concept.

6 Applications of Combined Cycles

The combined cycles described in chapter 5 are purely for electricity generation. In this chapter we explore how the thermodynamic advantages of combining gas and steam cycles can apply for other purposes—the production of thermal energy, or the repowering of existing steam power plants by integrating a gas cycle into the existing steam cycle. Some other, less conventional concepts are also discussed, although most of these have limited commercial usage.

Cogeneration

Cogeneration means the simultaneous production of electrical and thermal energy in the same power plant. The thermal energy is usually in the form of steam or hot water. The types of cogeneration plants discussed in this chapter fall into three main categories:

- Industrial power stations supplying heat to an industrial process
- District-heating power plants
- Power plants coupled to seawater desalination plants

The thermodynamic superiority of a combined-cycle plant over a conventional steam power plant is even more pronounced in cogeneration plants than in plants used to generate electricity alone. As shown in chapter 4, table 4–1, the temperature difference between the heat supplied to the cycle and the waste heat rejected to the environment is much bigger in a combined-cycle plant than in a steam power plant. This difference increases in relative terms if useful process heat is produced. In this case, the useful temperature drop gets reduced by the same amount, leading to an even higher ratio between combined-cycle plants and steam turbine plants.

A cogeneration plant may also have supplementary firing in the heat recovery steam generator (HRSG). This offers greater design and operating flexibility because the steam production can be controlled, independently of the electrical power output, by regulating the fuel input to the HRSG. The power output is controlled by the gas turbine. The cycle efficiency will, however, be normally lower if supplementary firing is used, not always making this an economically desirable solution.

Thermal energy in the form of steam can be extracted from the HRSG, the live-steam or reheat steam lines, or from an extraction in the steam turbine. The optimum extraction point depends on the required steam pressure, temperature, and quantity over the load range. These parameters are very important in defining the type of cycle that should be used. The advantages and disadvantages of various solutions are discussed in the following sections.

Industrial power stations

Wherever both electrical power and process steam are needed, it is thermodynamically and usually also economically better to produce both products in a single plant. The number of possible concepts is large because each plant is, to a certain extent, unique. The following examples have only one process steam supply. Often, steam supplies at different pressure levels are required, but the basic considerations remain unchanged. There are three basic steam extraction concepts that relate to the steam turbine:

- Cycle with back pressure steam turbine
- Cycle with extraction/condensing steam turbine
- Cycle with no steam turbine

The last concept is not really a combined-cycle power plant, but rather a gas turbine plant with HRSG.

A simplified flow diagram for a cycle with a back pressure steam turbine is shown in figure 6–1. The steam turbine (7) is designed so that the exhaust pressure matches the process requirement (3). A bypass (5) around the steam turbine is designed to reduce the live steam to the required pressure and temperature level for the process in case the steam turbine is out of operation and the process must still be supplied. In this cycle all of the steam generated in the HRSG goes into the process, except for a small quantity used for feedwater tank heating and deaeration.

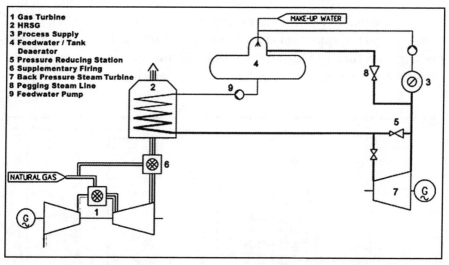

Figure 6–1 Simplified flow diagram of a cogeneration cycle with a back pressure turbine

For such a cycle, all or some of the steam supplied to the industrial process may be returned as condensate. Any loss in mass flow must be replaced with makeup water that usually enters the cycle at the deaerator because it is fully oxygenated. If condensate is returned from an industrial process, the quality may be poor and it may have to be treated before re-entering the water/steam cycle to achieve a suitable quality. If the return condensate is at a high temperature, this may influence the temperature at which the feedwater tank operates or the amount of heating steam to be extracted from the steam turbine to get sufficient deaeration and feedwater heating.

Figure 6–2 shows a concept using an extraction/condensing turbine (6). Here, the process steam is extracted from the steam turbine. The process steam system (5), which must be held at a constant pressure level, is regulated internally (4) within the steam turbine. A backup supply for the process is provided through the live-steam reduction station (7). The steam, which is not extracted for the process, further expands in the turbine and is condensed in a cold condenser (8), as for a noncogeneration plant. For such cycles, even without supplementary firing, the process steam flow can be varied without affecting the gas turbine load. This provides some additional operating flexibility, but not to the same extent as with a supplementary firing in the HRSG, which allows a steam output even at a low electrical output. This is because for the same gas turbine output, the steam turbine output and the process extraction flow are interdependent in a smaller range.

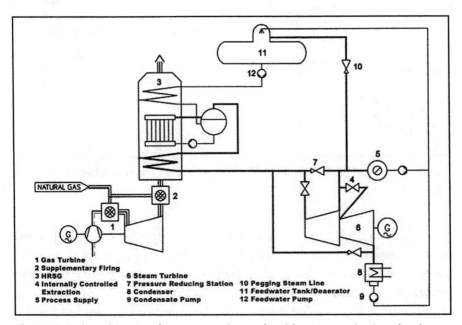

Figure 6–2 Flow diagram of a cogeneration cycle with an extraction/condensing steam turbine

A cycle with no steam turbine is shown in figure 6–3. This is appropriate if a large quantity of process steam is required at a high pressure and temperature, leaving little pressure drops for a steam turbine expansion. In this example, there is supplementary firing to achieve sufficient process steam mass flow. The steam goes directly from the HRSG to the industrial process (5) and the HRSG is designed to generate steam at the process conditions. The pressure may be controlled by a pressure reducing station (4), but it is quite common for the main pressure control to take place within the plant to which the process steam is being exported.

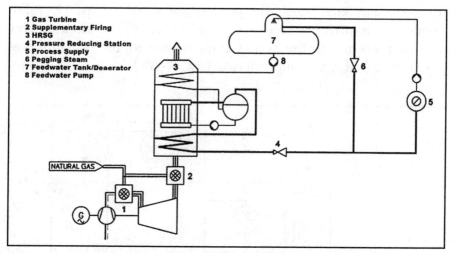

Figure 6–3 Flow diagram of a cogeneration cycle with no steam turbine

Evaluation of a cogeneration cycle

The performance of a cogeneration cycle is defined not only by the efficiency, but also by parameters such as fuel utilization and power coefficient. These parameters take into account the thermal as well as the electrical output.

Fuel utilization is a measure of how much of the fuel supplied is usefully used in the plant. It is equal to the sum of electrical output and thermal output divided by the fuel input.

The power coefficient (also called the alpha-value) is defined as the ratio between the electrical and the thermal output. Combined cycles tend to have high power coefficients; they are, therefore, more likely to be used for cogeneration applications with a relatively high power demand. This is because the electrical output of the gas turbine—about two-thirds of the total plant output—cannot be converted into thermal energy. In a conventional steam power plant, all of the energy produced could be exported to the process, giving a possible power coefficient of zero.

Figure 6–4 is a heat balance for a cycle with a back-pressure turbine using the 178 MW gas turbine and a single-pressure cycle with supplementary firing to 750°C (1382°F). The exhaust pressure of the steam turbine (and, hence, the process pressure) is 3 bar (29 psig). Condensate returns from the process at 50°C (122°F). A small amount of steam is taken from the steam turbine exhaust for feedwater heating to achieve a feedwater temperature of 60°C (140°F).

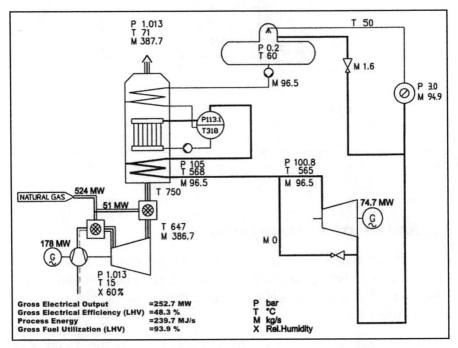

Figure 6–4 Heat balance for a single-pressure cogeneration cycle with supplementary firing

The cycle is directly comparable to figure 5–60 (single-pressure cycle with supplementary firing for electricity generation only), except that the cold condenser of figure 5–60 is replaced by the industrial process operating at a higher pressure level. Performance of the HRSG is the same in both cycles with identical stack temperatures and steam flows. Due to the high turbine back pressure, the output in the cogeneration cycle is 50.8 MW less than in the condensing cycle. The electrical efficiency decreases from 58% to 48.3%. However, 239.7 MJ/s of thermal energy is exported to the process in this case. This results in a fuel utilization of 93.9%. The remaining 6.1% of the energy is lost through the stack (5.5%), in radiation losses, and in generator and mechanical losses in the steam turbine and gas turbine. There is no condenser loss because all of the steam is used in the process. The power coefficient is 105%.

The power coefficient of a plant is mostly affected by three parameters:

- Amount of fuel supplied directly to the boiler (if any)
- Size of the condensing portion of an extraction/condensing turbine
- Pressure level of the process steam

Figure 6–5 shows the effect of process steam pressure on the relative power output and power coefficient for a combined cycle with a back-pressure steam turbine. The relative power output is the ratio between the electrical output of the considered cogeneration plant and the output of a condensing plant based on the same gas turbine. The reference point at 100% relative power output is equivalent to a condensing cycle. The higher the process steam pressure, the less electricity is produced because there is a smaller enthalpy drop in the steam turbine. In theory, the power co-efficient approaches 170% for back pressures equal to those of condensing applications. As the pressure increases the power coefficient falls exponentially to the limit set by the gas turbine output because the steam turbine electrical output is falling and the thermal output rising at the same time.

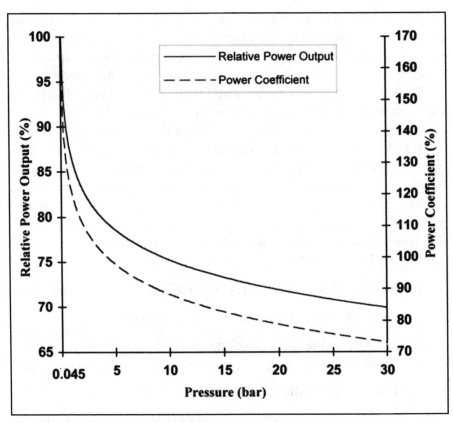

Figure 6–5 Effect of process steam pressure on relative combined-cycle power output and power coefficient for a single-pressure cycle with 750°C supplementary firing

As the power coefficient increases, the electrical efficiency approaches asymptotically that of a fully condensing combined-cycle plant (figure 6–6). The fuel utilization remains nearly constant because in a plant with no condenser, almost all of the steam entering the steam turbine is usefully used. With a variation of the process steam pressure, there is a shift between electrical and thermal output. The sum of both remains nearly constant except for some losses in the generator and transformer.

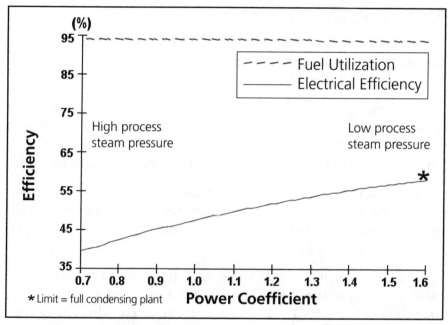

Figure 6–6 Effect of power coefficient on electrical efficiency and fuel utilization for a single-pressure cycle with 750°C supplementary firing

Where supplementary firing is installed, additional steam generation contributes to the electrical output of the steam turbine only and has no impact on the gas turbine output. Thermal output, therefore, increases at a much greater rate than electrical output with increasing supplementary firing temperature resulting in a decrease in the power coefficient.

If higher power coefficients are required, an extraction/condensing steam turbine offers greater design and operating flexibility. The condensing portion of the turbine enables the electrical power output to be increased, but only at the cost of thermal output and fuel utilization. If designed properly, such a cycle can operate at any point in the range from full condensing to full back-pressure mode. For back-pressure mode, cooling steam may be required for the low-pressure part of the steam turbine.

The highest fuel utilization can be obtained with a cycle with a back-pressure turbine but the flexibility is poor because the steam production can only be reduced by decreasing the gas turbine load. This measure would, however, have a negative impact on the electrical efficiency.

Main design parameters

The behavior and the influence of the design parameters of a combined-cycle power plant for cogeneration application is similar to the plants described in chapter 5 designed for pure power generation.

The biggest difference is because with increased process-steam pressure it may be better to increase the life steam pressure to keep a reasonable high enthalpy drop in the steam turbine. This is especially true when back-pressure steam turbines with a high back pressure are used.

However, this leads to poorer heat utilization in the HRSG in the case of a single-pressure process. To avoid these losses, a multipressure cycle can be used where the low-pressure steam can directly be fed into the process steam system.

Figure 6–7 shows a cycle using an unfired dual-pressure HRSG (2) and a back-pressure turbine (3). The LP steam is fed directly into the process steam system (4). The process steam supply is backed up with a pressure reducing station (6) from the HP live-steam line. It may be necessary to regulate the temperature of the process steam by injecting feedwater in the process steam line.

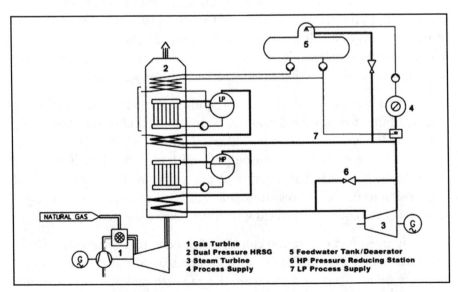

Figure 6–7 Flow diagram of a cogeneration cycle with a dual-pressure HRSG

District Heating Power Plants

A district heating system uses hot water or steam to supply heat for house heating. An efficient way to produce this hot water or steam is by extracting steam from a power plant at the required temperature level. A combined-cycle power plant is well suited for this type of cogeneration of power and heat.

District heating water is required at temperature levels much lower than those used for processes in industrial power plants. Usual forwarding temperatures are between 80 and 120°C (176 and 248°F) with return temperatures of 40 to 70°C (104 to 158°F).

Steam extracted from the LP part of the steam turbine is used to heat this water. For exergetic reasons in larger installations it is better to divide the heating into two or even three pressure stages. Figure 6–8 shows the temperatures of hot water and steam required in a one-stage and a three-stage process. Where a single stage is used, all steam must be extracted at the higher pressure level. This means there is less steam expanding in the steam turbine, so the steam turbine power output is lower than the one with a three-stage system. If only a small temperature increase in the district heating water is required, a single stage of heating may be good enough.

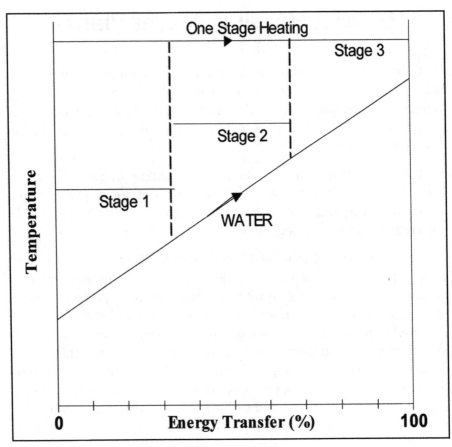

Figure 6–8 Comparison of one-stage and three-stage heating of district heating water

The main design criteria for district heating power plants are, therefore, the heat output and the heating water temperatures, according to the ambient temperature. The strong positive influence that low air temperature has on power output of the combined-cycle plant is an advantage in this case, as maximum heating output is demanded, when ambient temperatures are lowest. The design temperatures of the district heating water represent a compromise between maximization of electrical output and low costs for transportation of the heat.

The cycle must have a high degree of operating flexibility, but must not become too complicated or too expensive. In particular, district heating power plants should not be designed for extreme demand conditions because this would make them too expensive and bring into question the economical feasibility. Only about half of the peak heat demand will be covered by the cogeneration plant.

In district heating power plants, independent control of electricity and heat production is not always required. These power plants are usually integrated into large grids, where other power stations are available for adjusting the total electrical power output to meet the demand. Usually the only output parameter that must be controlled in the district heating plant is the thermal output and the water temperature.

Figure 6–9 shows the flow diagram and heat balance for a typical example of combined-cycle plant using an extraction/back-pressure turbine with heating in two stages and a dual-pressure cycle (with the same HRSG as figure 5–33). The first stage of district heating is done using the steam turbine exhaust steam. The exhaust steam, at a pressure of 0.38 bar (11.2"Hg), goes into the first district heating condenser where it is used to raise the temperature of the district heating water from 50 to 70°C (122 to 158°F). An extraction from the steam turbine at 0.78 bar (23"Hg) is used for the second stage of district heating, bringing water temperature up to 90°C (194°F).

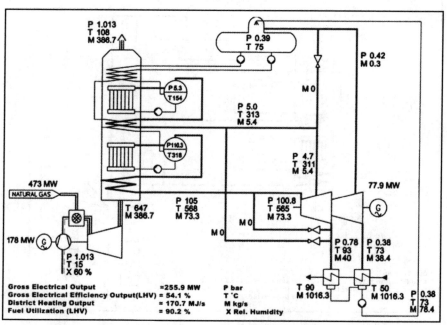

Figure 6–9 Heat balance for a cycle with two stages of district heating

Approximately half of the steam mass flow is used in each district heating condenser, and at the design point it is optimal to have the same heat load in each stage, as shown in this case. Condensate from the second-stage condenser cascades into the first-stage condenser and is returned to the feedwater tank.

Compared to the condensing cycle of figure 5–34, steam turbine output has fallen by 21.1 MW due to the steam extraction and the higher back pressure. The electrical efficiency falls from 58.6% to 54.1%, but the fuel utilization is reaching 90.2%.

Figure 6–10 shows a cycle using an extraction/condensing steam turbine. This time, each stage of district heating uses a steam turbine extraction, and there is a cold condenser at the exhaust of the steam turbine into which the steam not used for district heating is condensed after having been expanded in the low pressure steam turbine. This type of cycle can be used if there is a need to increase the flexibility of operation between district heating and electricity generation or in case a higher electrical output is required.

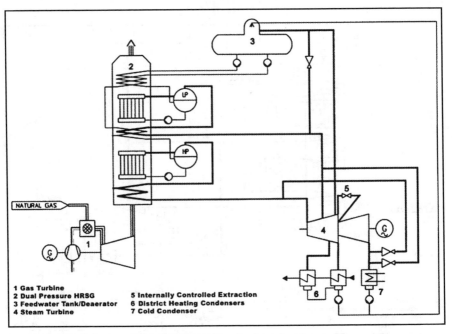

Figure 6–10 Flow diagram of a district heating/condensing cycle

Cycles coupled with seawater desalination units

In desalination plants, sweet water is produced out of salt water (usually seawater). There are several possible processes to desalt seawater, but often the process involves distillation, most of them use heat to evaporate part of the seawater. By condensation of the steam produced sweet water is obtained. A combined-cycle plant is a good source of heat for this evaporation process.

Combining a power station with a seawater desalination unit is particularly suited to combined-cycle plants because such power plants are generally built in oil- and natural gas-rich countries, where ideal fuels for combined cycles are readily available at a reasonable cost.

Larger desalination plants are usually designed to use the multistage multiflash process. Seawater is heated in a row of cells with increasing pressure with the final stage of heating (at around 100°C [212°F]) taking place in a heater supplied with steam from the combined cycle. Heated water returning through the cells heats incoming seawater in a counter flow heat exchange process. On entry to each cell the heated water undergoes a sudden pressure drop, causing instant evaporation of a part of the water, known as *flashing*. A part of the steam produced then condenses and is collected while the remainder recondenses, heating the cold water entering the cell before it moves on to the next stage. The heating steam is usually returned to the water/steam cycle in the form of condensate that—provided the quality is in order—can be admitted directly into the feedwater tank.

The maximum temperature of the water being heated is limited to prevent excessive formation of calcium carbonate ($CaCO_3$) deposits. This limit is between 90°C (194°F) and 120°C (248°F) depending upon the type of additives being used. Corresponding heating-steam pressures are between 1 and 2.5 bar (0 and 22 psig), which are ideal for a combined-cycle plant because a back-pressure steam turbine with a high enthalpy drop can be used, ensuring a high electrical output.

The electrical power output and the flow of process steam must usually be controlled independently of one another, so it is appropriate to install supplementary firing or to use an extraction-condensing steam turbine.

Figure 6–11 shows a typical flow diagram based on a single-pressure, supplementary-fired HRSG, a back-pressure turbine that feeds a multiflash desalination system.

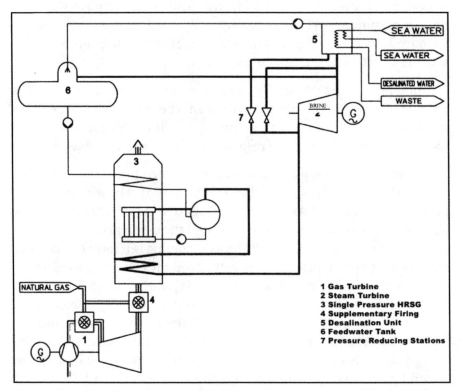

Figure 6–11 Flow diagram of a cycle coupled with a seawater desalination plant

Heating steam is directly exhausted from a back pressure turbine (2) to the desalination unit (5), where it is used to heat the seawater. The steam condenses in the desalination process and is returned to the feedwater tank (6). If the steam turbine is out of operation, supply of the desalination steam is provided through a pressure-reducing station (7). The high quantities of process steam at a low pressure require very large diameter pipes to the desalination unit.

With a combined cycle, electrical efficiencies of more than 50% can be reached for such desalination applications. A conventional steam power plant or gas turbine would not reach this level of efficiency.

A combined-cycle plant is less suitable if the ratio between fresh water produced and electrical power must be high, such as a low power coefficient. In that case, either the additional steam demanded for the desalination unit must be supplied from an auxiliary boiler, or a different type of power plant must be chosen.

Repowering

In this part various applications of the principle of combined cycle to repower existing steam or gas power plants are described.

Converting existing steam power plants into combined-cycle plants is known as repowering. It is ideal for plants in which the steam turbine, after many years of operation, still has considerable service life expectancy, but the boilers are ready for replacement. The boilers are normally replaced or supplemented with gas turbines and HRSGs. Steam turbine units in older power stations generally have relatively low live-steam data and can easily be adapted for use in a combined cycle. Repowering to a combined cycle can improve the efficiency of an existing plant to a level relatively close to that of new combined-cycle plants. Some plants are repowered purely to benefit from this efficiency increase, even though they are far from the end of their design life.

Conversion of conventional steam power plants

Figure 6–12 is a simplified flow diagram for a typical conventional steam turbine plant before repowering. The conventional boiler (1) generates steam for the nonreheat steam turbine (2). There are four extractions from the steam turbine for feedwater preheating–three to LP preheaters (4) and one to the feedwater tank (3). Condensing in this case is done using a water-cooled condenser (5). HP preheaters and reheat are not shown in this example, but their presence would not affect the principle of the repowering concept.

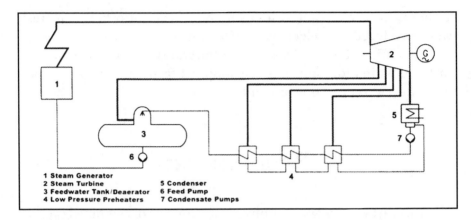

Figure 6–12 Flow diagram of a conventional nonreheat steam power plant

Figure 6–13 shows the same cycle after repowering with an oil-fired gas turbine and single-pressure HRSG. A preheating loop is installed in the HRSG to supply steam for deaeration (6). This means that the existing heaters are not needed, and the steam turbine extractions must be blocked off. An additional feedwater pump (15) must be installed for the preheating loop.

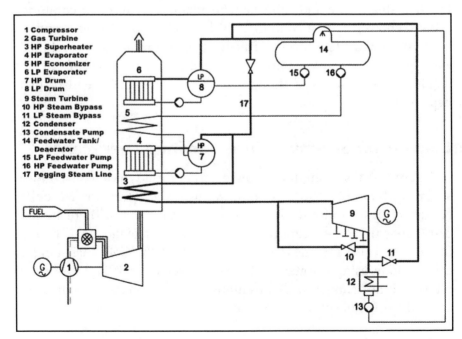

Figure 6–13 Flow diagram of a combined cycle plant using an existing steam turbine

For natural gas fuel with low sulfur content, a dual-pressure cycle would have been chosen leading to a better utilization of the exhaust gas.

If a concept using several pressure levels or even reheat is used, the question arises, where to feed the steam in the existing steam turbine. LP steam can often be admitted into the crossover pipe between the HP and LP parts of the turbine. If the steam turbine is too small, it may not have an internal crossover pipe. In such a case, one of the existing feed heating extractions could be adapted for use as an admission point.

When repowering, a decision must be made about which equipment will be retained from the existing steam plant. This will vary from case to case and depends on technical, economical, and reliability criteria. Typically, the following systems will be reused:

- Building and foundations
- Steam turbine and generator
- Condenser (maybe with new tubes)
- Main cooling system
- Main transformer for the steam turbine
- The high-voltage equipment

These larger components would be expensive to replace. It may, however, be preferable to replace most of the smaller components of the steam cycle because this can be done at relatively low cost. Retaining them may create operational problems and have a negative effect on the availability of the repowered installation. These items include:

- Condensate pumps
- Feedwater pumps
- Control equipment
- Piping and fittings
- Valves

To achieve efficiencies close to those of green field combined cycles it is important to have a good fit between the size of the gas turbine(s) and the steam turbine and that the steam turbine has live-steam parameters close to the optimum values of a new plant. If the steam turbine is too large and the gas turbine is too small, the HRSG will not be able to produce enough steam to build up sufficient steam pressure in the steam turbine to achieve good thermodynamic cycle data.

In case the steam turbine has lower live-steam data than required for an optimum cycle, the efficiency will also suffer.

Gas turbine with a conventional boiler

Modern steam power plants with reheat steam turbines can be repowered using a concept called the *fully fired combined-cycle plant* to improve the efficiency.

Conventional boilers use fresh-air fans to supply the air for combustion. Combustion air can also be provided by the exhaust gas of a gas turbine, if it is installed near the existing steam generator. The boiler must, however, be adapted to this new operating mode. Because the temperature of gas turbine exhaust is much higher than air after an air preheater–typically up to 550 to 650°C, (1022 to 1202°F) versus 300 to 350°C (572 to 662°F)–modifications of the burners, and maybe of some heat transfer sections of the boiler, are required. The high temperature of GT exhaust gas requires a quite sophisticated ducting system between the GT and the boiler.

Figure 6–14 is the flow diagram for such a cycle. The gas turbine exhaust gas flows first through the conventional boiler (4) where the remaining oxygen is used for the combustion and then through a waste-heat recovery system (5 and 8) used for most of the feedwater preheating. The rest of the preheating is done using the existing steam turbine preheaters (8 and 9). The fresh air fan (14) is retained for use in case the gas turbine is out of operation. The bypass stack in the gas turbine exhaust (3) provides extra operating flexibility and allows the gas turbine to operate in single-cycle mode if the water-/steam-cycle or main boiler is out of operation.

Such repowering is highly sophisticated and costly. The economical value of this solution is often questionable.

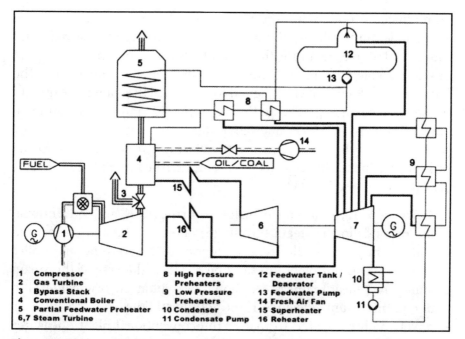

Figure 6–14 Flow diagram of a gas turbine combined with a conventional steam cycle (fully fired combined-cycle plant)

Obviously a key factor when considering such a conversion is the availability of space close to the boiler so that the gas turbine and the heat recovery system can be installed at the end of the boiler.

For steam power plants that burn gas or oil, the efficiency can be raised thanks to this conversion by more than 10% (relative) and power output by 20 to 30%. With coal-burning units, there is less potential gain because the conversion itself is even more complex and there is less improvement in efficiency.

Parallel-fired combined-cycle plant

Figure 6–15 shows the parallel-fired combined-cycle plant, a repowered cycle designed to have a very good efficiency combined with high operating flexibility.

A special aspect of this concept is that it is primarily intended for conventional coal fired plants where the conventional boiler is to remain unmodified in operation. The gas turbine (1) and HRSG (2 and 3) are installed in parallel to the conventional boiler to provide a second source of HP live steam for the steam turbine. Also here, the exhaust heat at the cold end of the HRSG is used for feedwater preheating. The HP section of the HRSG is designed to generate HP steam to be used in the steam turbine in parallel to the steam provided by the conventional boiler. The existing steam and feedwater heaters are reused but only partially fed with water, reducing the steam flow extracted from the turbine and increasing the electrical output. The remaining feedwater is heated up in the HRSG.

In Figure 6–15, total live-steam flow from the HRSG and conventional boiler expands in the steam turbine (5) and the total amount of cold reheat steam is fed to the reheat section of the conventional boiler where it is reheated before expanding through the rest of the steam turbine. The reheat temperature should be kept at the original level to maintain maximum efficiency. If the additional flow from the HRSG is relatively high compared to that of the conventional plant, measures must be taken to maintain the reheat temperature (e.g., exhaust gas recirculation, tilting burners, etc.).

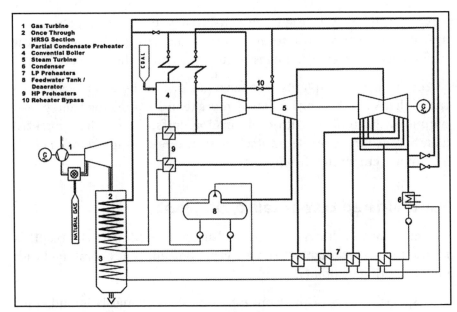

Figure 6–15 Flow diagram of a parallel-fired combined-cycle plant

There are three main operating modes for this concept:

- Original mode without the gas turbine and HRSG in operation
- Parallel mode, where the conventional cycle, gas turbine, and HRSG are in operation at the same time
- Pure combined-cycle mode, where the gas turbine, HRSG, and steam turbine are in operation without the conventional boiler. If the original cycle has reheat, a reheat bypass is necessary for this mode of operation.

Table 6–1 shows some performance data for a converted 500 MW conventional steam turbine power plant with the 178 MW gas turbine.

Table 6–1 Conversion of a 500 MW steam turbine power plant into a parallel-fired combined cycle plant

		Original mode	Parallel mode	Combined Cycle mode
Fuel Input (LHV)	MW	1,312.9	1,496.6	473.0
Steam Turbine Output	MW	502.3	495.4	87.0
Gas Turbine Output	MW	0	178	178
Total Gross Output	MW	502.3	673.4	265.0
Gross Efficiency (LHV)	%	38.26	45.0	56.0
Auxiliary Consumption	MW	32.7	30.5	4.9
Total Net Output	MW	469.6	642.9	260.1
Net Efficiency (LHV)	%	35.77	42.96	55.00
Net Heat Rate (LHV)	kJ/kWh	10,065	8,380	6,545
Net Heat Rate (LHV)	Btu/kWh	9,540	7,943	6,204
Net Marginal Efficiency of Gas (LHV)	%	–	58.90	55.00

Highest overall efficiencies are obtained with pure combined-cycle mode and the highest output with hybrid mode. However, in parallel mode the net marginal gas efficiency (the contribution of the gas-fired in the gas turbine to the total efficiency) is higher than in combined-cycle mode. This means that plant fired in parallel is a very efficient way of burning natural gas, making this an interesting alternative to separate plants. Combinations of the previously mentioned modes are also possible resulting in high operating flexibility and a wide operating regime. The conventional boiler can be fired with coal, oil, or gas.

The choice of operating mode at any moment will depend on the market prices for electricity and fuels. As the relative prices of gas and coal and the demand for electrical output vary, the operator can decide which mode can be run most economically at any one time.

It must, however, also be mentioned here that this solution is complicated and will not gain a large market acceptance.

Conversion of an Existing Gas Turbine Power Plant into Combined Cycle

The conversion of an existing simple-cycle gas turbine plant into combined cycle is quite easy compared to the conversion of a steam power plant. Downstream of existing gas turbines a steam cycle can be added, having the same characteristics as the steam cycle of an optimized new combined-cycle plant. The major issue is the availability of space close to the gas turbines to install the HRSG without long ducts for interconnection. The other factors to determine whether such a conversion makes economical sense are age and status of the gas turbine.

If the gas turbines are too old or in bad condition it does not make much sense to install a complete new steam part, which has much higher specific cost than the gas turbines (approximately twice the cost of a gas turbine plant).

In addition with older gas turbines, the resulting combined-cycle plants will only have a low efficiency compared to a completely new plant.

Special Applications

There are several variations in the way a gas cycle can be combined with a steam cycle, which can largely deviate from the type of combined cycles discussed so far. Some of these cycles have proven to be of some interest; others may gain some acceptance in the future; and some will never be realized. As yet, few of these have achieved any commercial status because they are either complex or costly or use components, which are not commercially available. Some of the most important of these applications will be discussed here.

Applications with alternative fuels

Where no natural gas is available, there is increasing interest in looking for ways to use available alternative fuels such as coal, heavy oil, and crude oils. Often these fuels have high sulfur content or are unsuitable for firing in gas turbines, so they must be treated before they can be used. Generally steam or heating is needed in the fuel treatment process, which can be provided by the combined-cycle plant and results in some very interesting but complex integrated processes. One of these processes has gained some level of commercial acceptance and is described in chapter 12.

Pressurized fluidized-bed combustion (PFBC). A typical cycle of this type is shown in figure 6–16. The gas turbine has an intercooler (4) between a low-pressure (1) and a high-pressure (2) compressor. At the outlet of the compressor, air is fed to the combustion vessel (5) with a pressure of 12 to 16 bar (160 to 218 psig) and forced through the fluidized bed where the combustion of the coal takes place. Coal is mixed with sorbent before entering the combustion process to bind chemically the sulfur contents in the coal. To obtain a high degree of desulfurization, the bed is maintained at a temperature of 850°C (1560°F). Before leaving the combustion vessel, the exhaust gas is led through cyclones (8) to clean it before it is expanded in the gas turbine (3). The exhaust heat of the gas turbine is recovered in an economizer (10), which is used for the final stages of feedwater preheating (17). The steam turbine is a single-pressure turbine with reheat (11).

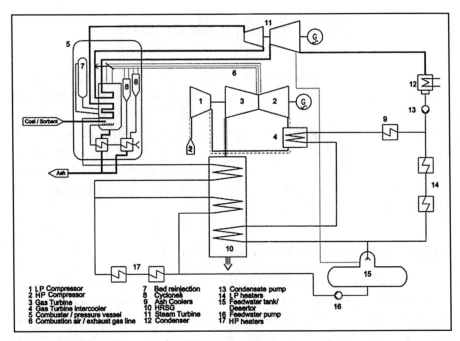

Figure 6–16 Flow diagram of a PFBC process

Net efficiencies are clearly above those of conventional power plants burning coal, although well below the level of pure combined-cycle plants because of the relatively low gas turbine inlet temperature (i.e., 850°C [1562°F]).

Another advantage of this cycle is that the fluidized bed with its steam generator is very compact. It takes much less space than a conventional boiler, although this is partly counteracted by the large economizer and the gas turbine. The small storage capacity of the steam generator means that the steam process reacts very quickly to changes in demand from the load control system.

However, the design is complex and requires a specially designed gas turbine very different from today's standardized industrial gas turbines. This means that the investment cost are very high compared to the normal gas-fired combined-cycle plant. Consequently, there is little chance that this system will ever gain market acceptance. The only major supplier having promoted this technology is ABB, who has left this business altogether.

Steam injection into the gas turbine (STIG) cycles

We have already seen how injection of steam into the gas turbine can be used to reduce NO_x emissions. This decreases the efficiency of the combined cycle, but increases the power output significantly. Cycles in which steam is injected into the gas turbine are known as steam-injected gas turbine (STIG) cycles.

Figure 6–17 shows a STIG cycle in which all the generated steam is directed into the gas turbine (4), apart from a small amount used for feedwater preheating and deaeration. Because steam is required only in the gas turbine at one pressure level, a single-pressure HRSG is used (2). The quantity of makeup water to the cycle is very high because all of the generated steam is lost to the atmosphere in the process. This type of cycle is suitable for use as a peaking unit in countries where water is plentiful. It is simple and cheap, attaining a high specific output and a higher efficiency than the gas turbine alone, although not as high as that of a combined cycle. However, if the injected steam flow is equal to more than approximately 2% to 4% of the air-mass flow, major modifications must be made to the gas turbine. This limits the commercial viability of the concept because gas

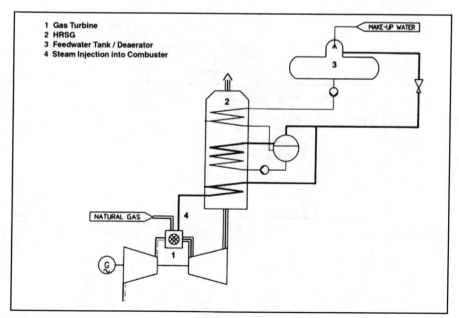

Figure 6–17 Flow diagram of a STIG cycle

turbines are standardized and are only modified for special applications when a substantial economic benefit can be expected.

Another, more efficient variation on this type of cycle is the Turbo STIG cycle, shown in figure 6–18. It is suitable for smaller plants with aero-derivative gas turbines. A steam turbine (4) is installed on the same shaft as the gas turbine (1) and shares the gas turbine generator. The single-pressure HRSG (2) generates steam for the steam turbine, expanding it to a level suitable for reheating in the HRSG and injecting (6) into the gas turbine. Thanks to the steam turbine, the power output and efficiency of this cycle is higher than the STIG cycle, but still does not reach the level of a normal combined cycle. This, and the fact that the water consumption is so high, is part of the reason for the limited commercial acceptance of the STIG cycles or Turbo STIG.

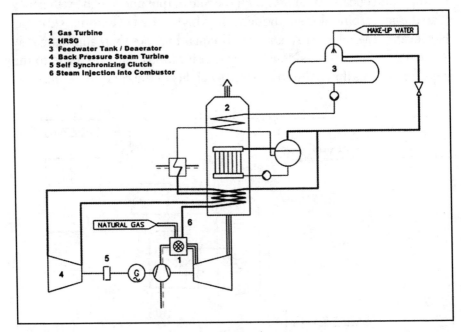

Figure 6–18 Flow diagram of a turbo STIG cycle

Humid air turbine (HAT) cycle

The principle here is to increase gas turbine output by using humid air as the working fluid. A flow diagram is shown in figure 6–19. Heat from compressor cooling (3 and 4) and the gas turbine exhaust is used to generate steam, which is being mixed with the compressed air. The result is an airflow with up to 25% steam content. This air is then heated, again using gas turbine exhaust heat, before entering the gas turbine.

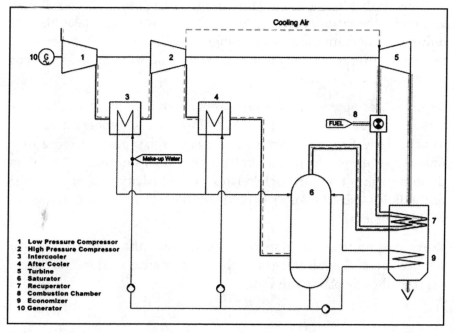

Figure 6–19 Flow diagram of a HAT cycle

Efficiencies reached with this cycle are not as high as those of normal combined cycles, but the HAT cycle can reach higher specific output levels. The cycle is also suited for use with coal gasification plants where available low-grade energy could be used to generate the required steam.

Difficulties in the HAT cycle lie in the operational complexity and the high water consumption that would cause problems where water is scarce. Also, new gas turbine development is needed due to the changed flow pattern in both the compressor and turbine of the gas turbine. This will probably be the main obstacle in commercializing this cycle.

Cycles with alternative working media

Another way of improving the performance of a combined-cycle plant is using fluids other than pure water/steam in the bottoming cycle. The idea is that by using a mixture of fluids, such as water mixed with ammonia, the evaporation in the HRSG will no longer take place at only one temperature but over a range of temperatures. This serves to reduce the exergy loss between exhaust gas and working fluid, thereby increasing the efficiency.

The cycles operate at only one pressure and are theoretically more suitable for gas turbines with relatively low exhaust temperatures, such as aero derivative machines. Condensation takes place at more than one temperature, which complicates the condensing system. Another disadvantage is that this cycle relates to the toxicity or aggressiveness of the fluids used. Leakages could lead to environmental pollution or health problems.

As gas turbines develop towards higher inlet and exhaust gas temperatures, the advantages of these alternative working media for combined-cycle plants will vanish.

7 Components

Gas Turbine

The gas turbine is the key component of a combined-cycle plant, generating approximately two-thirds of the total output.

The gas turbine process is simple: Ambient air is filtered, compressed to a pressure of 14 to 30 bar (190 to 420 psig), and used to burn the fuel, producing a hot gas with a temperature generally higher than 1,000°C (1832°F). This gas is expanded in a turbine that drives the compressor and generator. The expanded hot gas leaves the turbine at ambient pressure and at a temperature between 450 to 650°C (842 to 1202°F) depending on the gas turbine efficiency, pressure ratio, and turbine inlet temperature.

The combined-cycle plant has become a competitive thermal power plant only as a result of the rapid development in gas turbine engineering, which is still ongoing. Gas turbine development generally involves increasing gas turbine inlet temperatures (e.g., by improving cooling technologies) and increasing compressor air flows.

Increasing gas turbine inlet temperatures produce a higher useful enthalpy drop and, therefore, increase the efficiency and output of the gas turbine and the combined-cycle plant. This can usually be achieved using the same compressor, and even though an additional investment for materials may be necessary, the specific cost of the plant is reduced. Because fuel costs and capital costs are the main drivers for the cost of electricity generation, the gas turbine inlet temperature should be increased to the full potential of the material to improve the competitiveness of the product.

Parallel to the development of the turbine, there has also been some development in the compressor as well as substantial development in combustion systems. Today's compressors can handle much larger air mass flows and higher pressure ratios, resulting in higher power outputs, reduced specific costs, and improved efficiency.

Combustion technology plays a leading role in modern gas turbine technology. Rapidly growing demand for increased turbine inlet temperatures and tightened emission regulations represent a great challenge. Conflicting demands must be brought into harmony. The combustion system has to produce a turbine inlet temperature as high as possible for performance reasons, but maintain a flame temperature as low as possible due to emissions limits (NO_x production increases rapidly with flame temperature).

Stoichiometric combustion in diffusion burners results in high flame temperatures; therefore, the use of premix burners for lean combustion with significantly lower combustion temperatures compared to diffusion burners is common practice. Various design configurations exist for premix burners, but all are based on the targets of producing a uniform fuel/air mixture, operation with reliable flame stability (safeguarding against extinguishing and flame flashback), achieving uniform temperature distribution at turbine inlet, and, last but not least, ensuring complete combustion of fuel.

Conflicting requirements have to be overcome; these include aspects such as maximum air for the combustion process but also sufficient cooling air for hot gas path parts as gas temperatures increase. This results in smaller combustor designs with reduced surface area that requires less cooling air. One positive aspect that results from smaller combustors is the decrease in nitrogen oxides formation. NO_x formation directly depends on the residence time of hot flue gases in the combustor. On the other hand, complete burnout of fuel must still be ensured in these small combustors.

Such highly developed compact and highly stressed premix combustion systems tend to combustion-driven oscillations. Pressure fluctuations in the combustor driven by a thermo-acoustic resonance system causes pressure fluctuations at the burner outlet, which result in oscillating premixed fuel/air mass flow and additionally fluctuations in composition (fuel to air ratio) caused by differences in the acoustic properties of the air and fuel system. The resulting heat release fluctuations can create a self-induced feedback loop. Excessively increasing amplitudes in combination with the harmonic frequencies of the combustion system can cause combustion disturbances (flame flashback or extinction), mechanical influences, and pronounced casing vibrations resulting in wearing and fretting damage, and so on. In case high combustion dynamics occur during operation, it is good practice to use a protection system, which takes action on engine parameters to decrease the dynamics.

Additionally, competitiveness is increased with new gas turbine concepts, such as sequential combustion that appeared on the market in the 1940s and was re-introduced in the 1990s. The gas turbine has two combustion stages, with an intermediate turbine section, and a higher exhaust temperature. This means higher combined-cycle efficiencies can be achieved without raising the firing temperatures.

Figure 7–1 depicts the historical development of maximum air flows and gas turbine inlet temperatures.

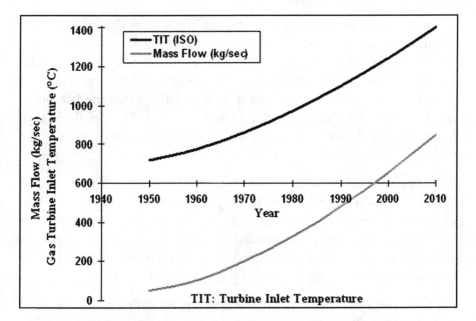

Figure 7–1 Historical trends in gas turbine inlet temperatures and compressor air flows

Two categories of turbines

Power generation gas turbines can be classified into two categories:

- The aeroderivative gas turbine, comprising a jet turbine modified for industrial duty and frequently incorporating a separate power turbine

- Heavy-duty industrial gas turbines, originally derived from steam turbine or jet technology

The *aeroderivative gas turbine* is normally a two- or three-shaft turbine with a variable-speed compressor and drive turbine. This is an advantage for part-load efficiency because airflow is reduced at lower speeds.

Inlet temperatures in jet engines are higher than in industrial gas turbines. In a jet turbine, weight plays the dominant role, and thus product and maintenance costs are less important than with industrial gas turbines, where long intervals between inspections are demanded. For that reason, and because of the smaller dimensions of the hot gas path, the trend toward higher inlet temperatures and greater power densities has progressed more rapidly in jet turbines than in stationary machines.

Aeroderivatives generally offer higher efficiencies than their frame counterparts as a result of aero technology. Furthermore, they are smaller and lighter for a given power output, and can be started rapidly because of their inherent low inertia. Because these turbines are derived from jet engines, they retain many of the features designed to allow rapid "on the wing" maintenance of aero engines. The small physical size of the aeroderivative enables it to be removed from its nacelle and replaced within a day.

A disadvantage when the aeroderivative is operating with a generator is that there is no compressor braking the power turbine in the event of load shedding. Two-shaft turbines are usually used for compressor or pump drives, where the operating speed of the power turbine is also variable. Due to the maximum size of aircraft, aeroderivatives are limited to approximately 50 MW electrical output.

Heavy-duty industrial gas turbines are practically always designed as single-shaft machines when used to drive generators with outputs greater than approximately 30 MW. Due to the increased turbine inlet temperatures and compressor air-mass flows, today's gas turbines for power generation can achieve electrical outputs of up to 340 MW. Figures 7–2 and 7–3 depict typical modern gas turbines. Figure 7–2 shows an aeroderivative unit, and figure 7–3 a heavy-uty industrial gas turbine designed for 340 MW output.

As gas turbines have become standardized, they have begun to be manufactured on the basis of sales forecasts rather than orders received. This results in a series of frame sizes for each gas turbine manufacturer, and, hence, shorter installation times, lower costs, and lower prices. Major developments in heavy-duty industrial gas turbines have been achieved in the last decade. A major impact has been made by the introduction to the market of gas turbines with sequential combustion.

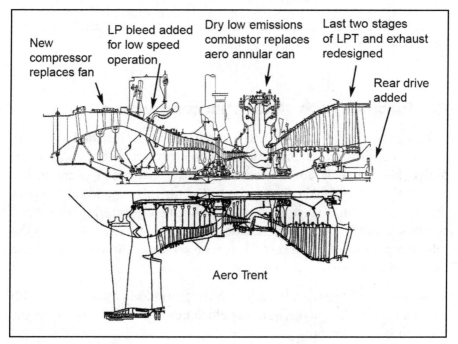

Figure 7–2 Industrial Trent derived from the aero Trent 800.
Source: Rolls-Royce plc. Registered trademark Rolls-Royce plc.

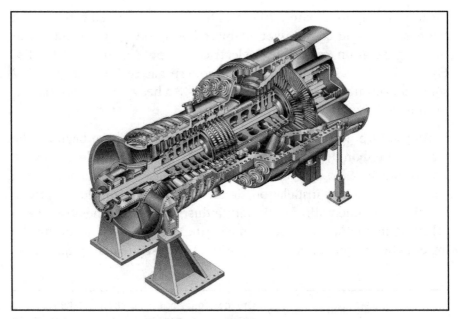

Figure 7–3 Heavy-duty industrial gas turbine

Gas turbines with sequential combustion

In a gas turbine with sequential combustion, compressed air enters the first combustion chamber located downstream of the compressor outlet. Here, fuel is combusted, raising the gas temperature to the inlet temperature for the first turbine. The hot gas expands as it passes through this turbine stage, generating power before entering the second combustion chamber, where additional fuel is combusted to reach the inlet temperature for the second turbine section, where the hot gas is expanded to atmospheric pressure.

With the same gas turbine inlet temperatures as in a gas turbine with a single combustion stage, a higher efficiency can be achieved with the same emissions levels.

Figure 7–4 shows a gas turbine with sequential combustion. The compressor is designed for a pressure ratio of 30, followed by the high-pressure combustion chamber, the high-pressure turbine, low-pressure combustion chamber, and low-pressure turbine. A compact arrangement fits all this equipment into one casing on a single-shaft.

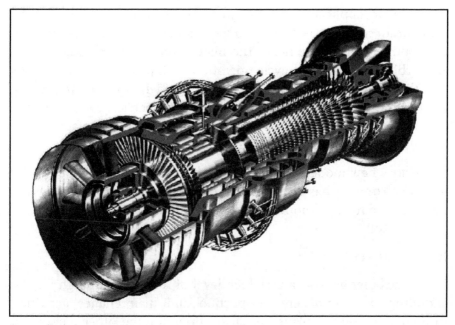

Figure 7–4 Gas turbine with sequential combustion

Table 7–1 lists the main characteristic data of modern gas turbines designed for power generation.

Table 7–1 Main characteristic data of modern gas turbines for power generation

Power Output (ISO condition)	up to 340 MW
Efficiency (ISO condition)	34–40%
Gas turbine inlet temperature (ISO 2314)	1,100–1,350°C (2,012–2,462°F)
Exhaust gas temperature	450–650°C (842–1,202°F)
Exhaust gas flow	50–820 kg/s (397,000–6,500,000lb/hr)

In the past, gas turbines were mainly developed for use in simple-cycle operation and the majority of gas turbine installations were of this type. These gas turbines were often used as peaking and stand-by machines. The first combined-cycle power plants were based on the same gas turbines, but operated in intermediate and base load applications.

The main application for the gas turbine is in a combined-cycle configuration; this has had a corresponding impact on gas turbine development and equipment. The main market breakthrough of the gas turbine in combined-cycle operation took place in the mid-1980s and the technology is now mature and widely applied. Improving gas turbine performance lowers the cost of electricity, increases the competitiveness of the product, and thus motivates manufacturers to invest in new designs and upgrades.

Because new models of given gas turbines are often sold before the first unit running has accumulated many operating hours, attention must be paid to corresponding risk mitigation measures.

Inspections

Gas turbines achieve a very high level of reliability through regular and properly administered inspection and maintenance activities. In a combined-cycle plant, the inspections for the other equipment can usually be completed in the window defined by the gas turbine inspection schedule.

Degradation

The output of a gas turbine is subject to degradation that has two main causes:

- *Fouling,* which is recoverable. During operation, attention must be paid to compressor and turbine fouling.

- *Aging,* which is non-recoverable unless parts are replaced.

Compressor fouling occurs because the gas turbine operates in an open cycle. The compressor ingests air that cannot be completely cleaned. Compressor fouling is reduced by an air filtration system that is suited to the environmental conditions at the plant site. The filters most frequently used are two-stage filters or self-cleaning pulse filters. The latter of these is only suitable for use in dry climates.

For economical operation of a power plant with gas turbines, proper design and size of the air filter system, together with well balanced maintenance intervals for compressor cleaning (on-line and off-line) and filter replacement, are essential. Pressure drops in the air intake system result in performance losses. One mbar intake pressure drop causes:

- Approximately 0.1% output loss
- Approximately 0.03% (approx. 0.013% point) efficiency loss

The design and size of the filter structure dictate the basic pressure drop of the intake system. The design pressure drop of a new and clean filter bank is normally on the order of 2 mbar. The approach velocity of air upstream of the filter surface is approximately 3 m/s.

It is, however, impossible to keep the compressor completely clean. The fouling that results causes losses in output and efficiency.

Two types of compressor cleaning can be used to help recover these losses:

- On-line washing
- Off-line washing

On-line washing is straightforward but because the temperature increases through the compressor, the cleaning solution evaporates and cleaning is limited to the first compressor rows.

Off-line washing is better suited to large, modern gas turbines because it is more effective. However, it requires shutting down and cooling the engine. Washing at low speed is preferred (e.g., ignition speed with the machine cold). The machine is, therefore, out of operation for approximately 24 hours, chiefly for cooling down prior to and for drying the engine after washing. This type of washing is best done before or after an inspection. The amount of cleaning solution needed is on the order of 40 to 200 liters per washing cycle.

Keeping degradation as low as possible requires an optimized maintenance program that includes a well-balanced combination of following measures:

- On-line compressor cleaning
- Off-line compressor cleaning
- Air intake filter replacement (partial and complete replacement dependent on distribution of dirt deposits)

A typical recommendation for replacement of air filter elements is a total pressure drop of 7 to 8 mbar over all filter elements.

After compressor cleaning in a gas-fired machine, any remaining degradation is due to aging. The symptoms of aging after 8000 hours in operation on a clean fuel are a:

- Reduction in combined-cycle plant output of about 0.8 to 1.5%
- Reduction in combined-cycle plant efficiency of about 0.5 to 0.8%

The rate of aging decreases with time between major overhauls.

Turbine fouling is caused principally by the ash formed by combustion of heavy fuels and in the additives used to inhibit high-temperature corrosion. It is unavoidable with heavy fuels, but can be limited by selecting the proper type of additives. Heavy- and crude-oil-fired gas turbines can also be equipped with a turbine washing system. Turbine fouling is less of a problem in peaking or intermediate-load operation because of the self-cleaning effect of startup and shutdown.

Typical degradation (after compressor washing) in combined-cycle plants after 8,000 hours of operation on heavy or crude oil are a:

- Reduction in combined-cycle plant output of 4 to 5.5%
- Reduction in combined-cycle plant efficiency of 1.5 to 1.9%

The rate of degradation also decreases with time. Figure 7–5 shows the correlation between degradation and maintenance measures with a simplified graph.

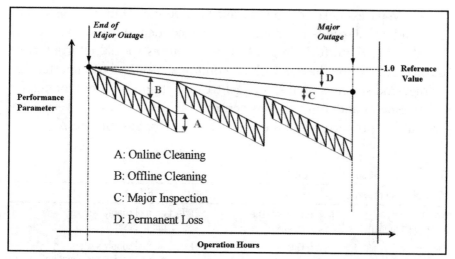

Figure 7–5 Types of losses contributing to overall performance degradation (source: ASME Paper 2003-GT-38184)

In the past, corrosion problems were one of the major causes of gas turbine failures. Because of the use of better blading materials and coatings, problems of this type have practically been eliminated. Whenever heavy fuels are used, particularly those containing vanadium or sodium, it is necessary to use additives or treat the fuel to prevent high-temperature corrosion. The additives commonly used are based on magnesium, chromium, or silicon.

Fuels for gas turbines

More than three-quarters of installed gas turbine capacity use natural gas and light distillate fuel whereas fuel oil is most often used only as a backup fuel. These clean fuels enable the highest performance with state-of-the-art gas turbines, allow clean combustion with the lowest emissions levels, are easy to handle, are compatible with approved standard combustors and auxiliary equipment, provide the simplest mode of operation, are synonymous with high plant availability, and require the lowest maintenance effort.

The remaining quarter uses unconventional fuels, such as special fuels like low BTU natural gas, off-gases from iron making processes, synthesis gases (syngases) derived from various source fuels, but also heavy fuel oil, residual oil, crude oil, naphtha or gas condensate fuels, and so on. Combusting these fuels in gas turbines requires special efforts for the burner design, for the gas turbine compressor/turbine design, for fuel storage and supply systems, for fuel treatment and/or conditioning, for safety aspects and for emissions control. Table 7–2 gives an overview of critical fuel properties and their consequences for gas turbines.

Table 7–2 Critical fuel properties

Critical Fuel Properties	Fuels	Effect
Low Viscosity (reduced lubricity)	Naphtha, Kerosene, Condensates	Effect on fuel supply system (e.g. pump design)
High Viscosity	Heavy Oils, Heavy Residues	Effect on fuel supply system (e.g. pump design, trace heating) Requires start-up, shut down and flushing systems Effect on fuel atomization (fuel heating)
Low density (high volume flow)	Naphtha, Condensates	Limits for fuel supply system and burner nozzles
Low flash point Low boiling point (high vapour pressure)	Naphtha, Kerosene, Liquified Petroleum Gases (LPG) High Speed Diesel (HSD) Condensates	Increased explosion protection effort (inkl. start-up and shut-down fuel system) Increased ventilation effort
Low ignition point (auto ignition)	Naphtha, Condensates	Effect on premix capability Increased explosion protection effort Increased ventilation effort
Contaminants (high temperature corrosion, ash deposits)	Contaminated fuel oils, Crude oils, Heavy Oils Heavy residues	Effect on blading and hot gas path. Counter-measures: - Temperature reduction (reduced performance) - Fuel treatment (washing/inhibitor dosing)
Low heating value (high volume flow)	Process and synthesis gases (low calorific gases), Low BTU natural gas	Effect on layout of compressor, burner nozzles, fuel supply system
High heating value (low volume flow)	LPG, gaseous and liquid	Effect on layout of burners, fuel supply system Limited start-up and part-load capability
High H_2 content	Process and synthesis gases (low calorific gases)	High flame velocity, Effect on premix capability more explosive than natural gas
High dew point	Gases with high boiling components	Droplets causes erosion and non constant heat flow

Gaseous fuels

For comparison of gaseous fuels the heating value is not the only criterion. The Wobbe index (or Wobbe number) is used for classification of fuels. The Wobbe index is a key parameter for the heat capacity of a gaseous fuel or, in other words, it is an indicator for the interchangeability of gases. Gases with same Wobbe index produce an identical burner heat load at the same burner pressure and can therefore be used in the same burner.

The Wobbe index for lower heating value is defined as:

$$W_i = \frac{Q}{\sqrt{\frac{\rho}{\rho_D}}} \quad [J/m^3, kWh/m^3 \text{ or } Btu/ft^3] \tag{7.1}$$

where W_i = Wobbe index, Q = heat input based on lower heating value, ρ = density of fuel gas at standard conditions, ρ_D = density of air at standard conditions.

Gases with same Wobbe Index or within a range of ±2 to ±5% for premix burners (±15% for diffusion burners) can be used in the same burner. The Wobbe index of a gaseous fuel can be adjusted by diluting it with inert or lean gases (e.g., steam, nitrogen) or improved by adding rich gases (e.g., evaporated LNG).

Some fuels are usually known as liquids, but will normally be vaporized for combustion in gas turbine combustors. Propane, butane, and liquefied petroleum gas (LPG) are typical fuels of this kind.

Natural gas in pipeline quality is the typical standard fuel suitable for use in standard fuel supply systems and combustors. Critical properties such as Wobbe index, heating value, composition, and the compounds with usually restricted concentrations such as hydrogen, higher hydrocarbons (>C3), hydrogen sulfide, and mercaptans are set and maintained within defined and acceptable limits.

Depending on pipeline pressure and the pressure ratio of the gas turbine, two different cases are possible:

- The pipeline pressure is too low and compression is necessary. In this case it is of essential importance to avoid any carryover of liquids (e.g., lubricants) from compressor equipment into the fuel gas stream.

- The pipeline pressure is too high and pressure reduction is necessary. The resulting temperature drop involves a risk of condensation of gaseous constituents. Separation of liquid phases and heating of the gas may be necessary.

In both of these cases, the control system for regulating the pressure of fuel gas supplied to the gas turbine system requires particular attention. Rapid flow changes are frequent, especially during gas turbine startup and shutdown, or, as an extreme case, on gas turbine trip. Typical gas pressure controllers are not able to handle such rapid changes with a fast response. Gas turbine controllers require a defined, constant fuel gas pressure upstream of the load control valve. Pressure fluctuations result in load fluctuations. To moderate such pressure fluctuations especially in multi-unit configurations, a certain buffer volume in the fuel gas supply system between primary pressure controller and gas turbine load controller should be provided.

The fuel gas temperature upstream of burners has to be kept in a certain safety margin above the dew points of water and H_2S to avoid formation of droplets in the fuel gas stream and ingress into the burners as well as to prevent corrosion.

Liquefied natural gas (LNG) has to be vaporized to a gas. Produced by a liquefaction process, LNG is a very clean and rich fuel gas, from which all higher hydrocarbons (C6+), inert components (N_2 and CO_2) and most impurities have been removed. Pure evaporated LNG has a higher Wobbe index than pipeline natural gas. Nevertheless, variations in fuel characteristics are encountered with LNG from various sources. Sudden changes in fuel characteristic must be avoided (e.g., between different shipments). In particular the advanced premix burners demand tighter control of the variability in fuel constituents.

Low BTU natural gas contains a higher percentage of nitrogen and carbon dioxide as well as impurities such as hydrogen sulfide and mercaptans. For calorific values below an engine-specific typical minimum value (in the range of approx. 35 to 45 MJ/kg), dry low-NO_x premix operation is excluded and for even lower values, modified burners and gas turbine compressors are required. Blending with other high-calorific value gases may be necessary to achieve levels necessary for ignition, flammability and Wobbe index as well as for composition of a constant, defined quality.

Off gases from iron and steel industries, such as top gas, blast furnace gas, and converter gas (COREX gas and coke oven gas are more similar to syngases and are listed there) range in the very low to medium BTU classes and have a high carbon monoxide content. They require careful cleaning that may involve removal of many inadmissible contaminants. Blending with high calorific gases as well as storage systems for intermittent processes (e.g., converter gas) are required to maintain a constant gas composition.

Syngases are mainly gasifier gases derived from various source fuels such as coal, biomass and refinery residuals, and so on. COREX gas and coke oven gas from iron production are included among the syngases due to their high hydrogen contents.

Typical characteristics of syngases are low calorific value, with high contents of hydrogen and carbon monoxide; therefore they require a special fuel supply system, modified burners as well as an adapted gas turbine compressor design. Furthermore, increased explosion protection measures and provisions to safeguard against toxic carbon monoxide must be implemented.

Modified burners are necessary for hydrogen rich fuels and increased mass flow to compensate for low calorific value. The higher flame temperature when firing hydrogen produces more NO_x than combustion of natural gas. Other concerns are related to the higher flame velocity of hydrogen and thus the risk of flashback in dry low-NO_x premix burners. Use of diffusion burners and NO_x control by diluting the syngas with inert gases (e.g., nitrogen) is commonplace.

Compressor modification is required by the low calorific value of these gases and the resulting higher turbine mass flows and therefore increased pressure ratio in the gas turbine.

Vaporized liquids such as propane, butane, or mixtures of the two (i.e., LPG) are low-boiling hydrocarbons that have to be vaporized for use in gas turbines. These fuels belong to the very high calorific value class. They are normally clean, but require a modified burner design and safety measures pertaining to explosion hazards.

When these fuels are used, it is essential that no liquid phase occurs in the fuel supply system that would result in disastrous load fluctuations of the gas turbine. The fuel system needs effective measures to prevent condensation of gas inside the piping system that occurs especially after shutdown and cool down of the system.

Table 7–3 lists typical compositions of fuel gases for gas turbines.

Table 7–3 Typical composition of fuel gases for gas turbines

		Natural gas Standard	Natural gas Low-BTU	Gasification coal	Gasification residue	Gasification biomass	Blastfurnace gas	Converter gas	COREX gas
H_2	[%vol.]		0.2	12.3	27.2	11.2	2.5	1.0	16.1
CO	[%vol.]			24.8	31.8	20.2	22.8	69.2	43.0
CO_2	[%vol.]	0.2	0.5	0.8	4.9	12.0	21.2	14.6	36.5
N_2	[%vol.]	3.0	62.3	42.0	0.5	44.6	53.5	14.9	2.8
CH_4	[%vol.]	93.0	36.0		0.3	5.6			1.6
Ar	[%vol.]			0.6	0.7				
H_2O	[%vol.]			19.1	34.6				
O_2	[%vol.]			0.4					
CnHm	[%vol.]	3.8	1.0			0.2		0.3	
Density	[kg/m³]	0.77	1.06	1.03	0.82	1.10	1.42	1.26	1.32
LHV	[MJ/kg]	47.0	12.8	4.3	8.6	5.3	2.3	6.7	5.9

Liquid fuels

Light distillates, according to defined standards as ASTM D2880 (Distillate No. 2-GT) or EN DIN 51603 (*Heizöl EL*), are products with fixed, reproducible properties that can be used with standard fuel supply systems, combustors, and hot gas path parts. Essential properties such as viscosity, density, ash, flashpoint, vapor pressure, admissible trace metal contamination levels (mainly sodium, potassium, calcium, vanadium, and lead), and so on are specified so that such fuels can be combusted without further measures and without any impact on normal performance levels, maintenance intervals, duration, and effort as well as the service life of parts.

Ash-forming fuels, such as heavy oils, residual oils, and crude oils, are generally contaminated with corrosive impurities (trace metals, mainly sodium, potassium and vanadium). These fuels require treatment before combustion. This includes heating to lower the viscosity for pumping, washing, and finally for atomization by the burner nozzles. Depending on the type and amount of contaminants present, turbine inlet temperature must be lowered (loss of performance), overhaul intervals are shortened (loss of availability), and hot gas path part replacement costs increase. The burner system is normally identical to that for light distillate.

Sodium, potassium, and other water-soluble contaminants have to be washed out by forming an oil-and-water mixture, and removed by centrifuging or electrostatic separation.

Oil-soluble contaminants, mainly vanadium but also lead, must be controlled with suitable inhibitors. Additionally, the gas turbine inlet temperature has to be lowered to keep it below the melting point of compounds formed from these contaminants to prevent molten corrosive products from adhering to the surface of hot gas path parts.

Maximum acceptable limits for these contaminants need careful monitoring to prevent the hot gas path parts from high-temperature corrosion, but also burner nozzles and the entire fuel system from wear, coking, and other forms of damage.

Unavoidable ash formation and the resulting deposits on blades and the entire gas flow path lowers performance in addition to the lower performance already caused by the reduced turbine inlet temperature. The requisite periodic off-line turbine washing after a minimum cool-down time of hot parts degrades the availability of the power plant; this is compounded by increased maintenance efforts associated with shorter intervals for overhauls and longer outages. More frequent replacement of hot gas path parts compared to plants operated on light distillate fuel has to be expected.

Heavy fuels normally cannot be ignited for gas turbine startup; therefore a startup and shutdown fuel—usually light distillate—is needed with its own storage, forwarding system, and fuel changeover equipment. This system must be able to flush the fuel system after gas turbine trip during operation on heavy fuel oil if flushing the pipes was not possible by normal fuel changeover. Thermal insulation of all fuel equipment as well as trace heating for piping sections that are not flushable also have to be considered.

Additional special features are valid for these ash-forming heavy fuels if they contain high volatile fractions. This applies especially for crude oil, but also sometimes for heavy oils or residual oils if they are blended. Necessary measures are described under low boiling and high volatile liquid fuels.

Low boiling and **high volatile liquid fuels** such as kerosene, naphtha (as a refinery distillation product), and gas condensates (as natural gas liquids) are mostly free from impurities (contaminants typically enter the fuel due to negligence during transport, storage and handling). Naphtha and gas condensates have a high calorific value (>42 MJ/kg) but a low density, low viscosity, and a comparatively high vapor pressure (max. 0.9 bar at 38°C compared to 0.003 bar of light distillate).

Such fuels have nearly similar operating conditions as light distillates in terms of combustion. However, safety requirements for storage, forwarding, and injection systems as well as for the environs are very extensive compared to light distillate or natural gas. Leak-tight systems, advanced ventilation, additional enclosures, leakage detection, and, in some cases, explosion-proof equipment are mandatory.

Fuels with high vapor pressures (e.g., naphtha) require a safe startup and shutdown fuel (normally light distillate) to prevent explosive vapors from entering the combustion chamber and ultimately the entire gas turbine.

The low viscosity and lubrication properties of these fuels require the use of centrifugal pumps instead of positive displacement pumps. Depending on the type of mechanical parts employed in the fuel distribution system, lubricating additives may also be necessary.

Inerting and flushing systems with safe media may be needed for components with open liquid surfaces (e.g., leakage collecting tanks) and for maintenance activities. If vapors are heavier than air, sewer systems as well as pipe and cable trenches have to be protected against vapor ingress.

These special features are also valid for liquid fuels with high volatile fractions (especially crude oil and lighter fractions blended heavy oils or residual oils).

Methanol is comparable to naphtha in terms of its basic requirements for fuel supply system and safety aspects (additionally, the toxicity of methanol has to be considered).

A significant difference to such other liquid fuels is methanol's low calorific value of approx. 20 MJ/kg, which is only about half that of light fuel oil or naphtha. Consequently, specific burners are necessary for methanol firing as well as an enlarged capacity fuel supply system.

Heat Recovery Steam Generator

The heat recovery steam generator (HRSG) is the link between the gas turbine and the steam turbine process. Three main configurations are widespread:

- HRSG without supplementary firing
- HRSG with supplementary firing
- Steam generators with maximum supplementary firing

As elucidated in chapter 5, HRSGs without supplementary firing are most common in combined-cycle plants.

The function of the HRSG is to convert the thermal energy in the gas turbine exhaust into steam. After heating in the economizer, water enters the drum at slightly subcooled conditions. From the drum, it is circulated to the evaporator and returns to the drum as a water/steam mixture where water and steam are separated. Saturated steam leaves the drum and is forwarded to the superheater where it is exposed to the maximum heat exchange temperature of the hottest exhaust leaving the gas turbine. The heat exchange in an HRSG can take place on up to three pressure levels, depending on the desired amount of energy and exergy to be recovered. Today, steam generation at two or three pressure levels is most commonly used.

HRSG without supplementary firing

Construction. A HRSG without supplementary firing is essentially an entirely convective heat exchanger. The requirements imposed by operation of the combined-cycle power plant are often underestimated. In particular, provision must be made to accommodate the short startup time of the gas turbine and the requirements imposed by quick load changes. The HRSG must be designed for high reliability and availability. HRSGs can be built in two basic configurations, based on the direction of gas turbine exhaust flow through the boiler.

Vertical HRSG. In the past, vertical HRSGs were most often known as forced-circulation HRSGs because of the use of circulating pumps to provide positive circulation of boiler water through the evaporator sections. In this type of boiler, the heat transfer tubes are horizontal, suspended from uncooled tube supports located in the gas path. Vertical HRSGs can also be designed with evaporators that function without the use of circulating pumps. However, whereas natural-circulation in horizontal boilers starts automatically, it must be ensured in vertical boilers by specific design measures.

Figure 7–6 shows a forced-circulation HRSG. The exhaust-gas flow is vertical, with horizontal tube bundles suspended in the steel structure. Circulating pumps ensure constant circulation within the evaporator. The structural steel frame of the steam generator supports the drums.

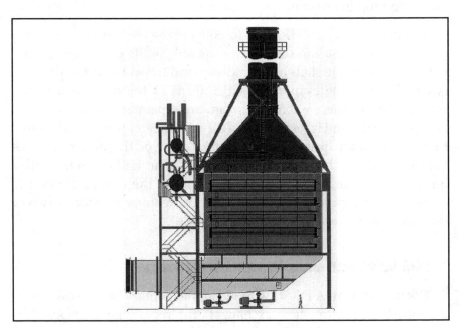

Figure 7–6 Forced circulation heat recovery steam generator

Horizontal HRSG. The horizontal HRSG is typically also known as a natural-circulation HRSG because circulation through the evaporator takes place entirely by gravity, based on the density difference of water and boiling water mixtures. In this type of boiler, the heat transfer tubes are vertical, and essentially self-supporting.

Figure 7–7 shows a natural-circulation HRSG. The exhaust-gas flow is horizontal. The steel structure is more compact than on a unit with vertical gas flow.

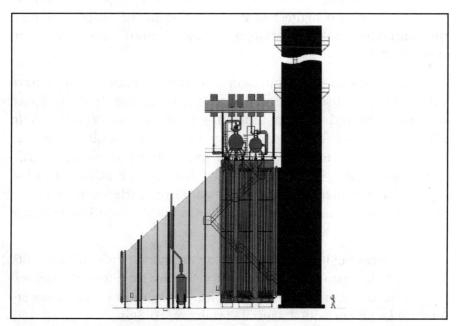

Figure 7–7 Natural circulation heat recovery steam generator

Design comparison. Either type—vertical or horizontal—can be used in a combined-cycle plant.

In the past, vertical HRSGs had several advantages that made them especially well suited to combined-cycle applications:

- Minimum footprint requirement arising from the vertical design

- Smaller boiler volumes because of the use of smaller diameter tubes
- Less sensitivity to steam blockage in economizers during startup

The main advantage of the horizontal HRSG is that no circulation pumps whatsoever are needed—an important consideration for applications with design pressures above 100 bar (1430 psig), where pumps must be designed and operated with special care. Additionally, there is an advantage with vertical tubes in the evaporator because the tubes with the highest heat absorption in the evaporator have the most vigorous circulation, and tube dry-out cannot occur in vertical tubes.

Current design techniques for natural-circulation boilers have overcome the disadvantages relative to vertical boilers. Space requirements and startup times are identical, water volumes in evaporators have been reduced by the use of smaller diameter tubes, and steam blockage is better handled in a modern, natural-circulation boiler. The same pinch points can be achieved in the high- and intermediate-pressure evaporator. Differences occur in steam performance only in large HRSGs with a tight low-pressure pinch point.

The presence of both technologies on the market indicates that both meet customer expectations and the preference for one or the other is more a historical or regional preference. Today most of the plants are equipped with horizontal natural-circulation HRSGs.

Once-through HRSG. The HRSGs previously described use a steam drum for water/steam separation and water retention. Combined-cycle plants are often operated in cycling duties with frequent load changes and start-stop cycles. HRSGs employing a drum with design pressures of 100 bar (1430 psig) and beyond impose restrictions on this mode of operation.

In a once-through steam generator, the economizer, evaporator, and superheater are basically one tube—water enters at one end and steam leaves at the other end, eliminating the drum and the circulation pumps. This design has advantages at higher steam pressures because the drum does not impose limits during startups and load changes. Both horizontal and vertical HRSGs can be built with the once-through circulation principle. The principle of once-through evaporation (BENSON® Once-Through design) compared to a drum-type boiler design is illustrated in figure 7–8.

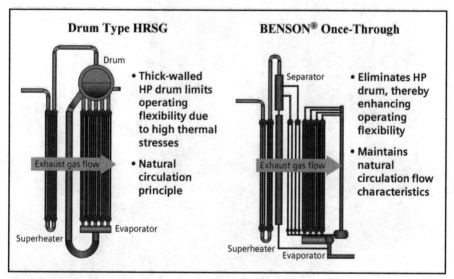

Figure 7–8 Principle of drum type and once-through evaporation

Advanced HRSGs for maximum efficiency are of a three-pressure reheat configuration with the high-pressure section in a once-through design, and medium- and low-pressure sections as drum-type evaporators. Figure 7–9 shows a three-pressure reheat once-through HRSG with horizontal gas flow. The high pressure section is of the once-through type, while the low and intermediate pressure sections are of conventional, natural-circulation design with LP and IP drums located on top of the boiler.

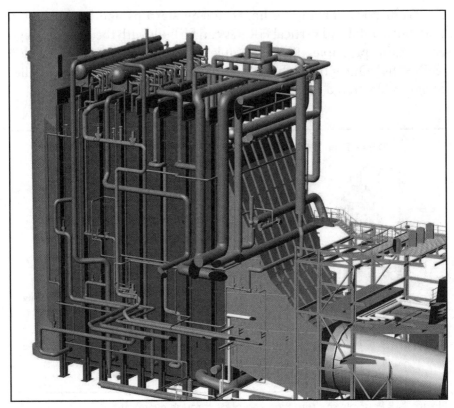

Figure 7–9 Three-pressure reheat once-through HRSG with drum type LP and IP sections

Casing of the HRSG. Two main variants are used for the casing of the HRSG:

- Cold-casing design with internal insulation
- Hot-casing design with external insulation

The cold-casing design is widely used with natural-circulation steam generators and has advantages at high exhaust temperatures. The casing withstands the exhaust pressure forces and the internal insulation keeps it at a low temperature. The casing construction also imposes no limits on startup time to allow for thermal expansion of the casing.

The hot-casing design is often used with the vertical HRSGs and has advantages at lower exhaust temperatures. When fuels with high sulfur content are fired, a hot casing at the cold end of the HRSG can limit corrosion.

With modern gas turbines running at high exhaust temperatures, a hot-casing design requires high-alloy materials in the hot end of the boiler. Today, some vertical HRSGs also use a cold-casing design.

Finned Tubing. Heat transfer in the HRSG is mainly by convection. Heat transfer on the water side is much better than on the exhaust side, and so finned tubes are employed on the exhaust-gas side to increase the heat transfer surface.

Normal fin density for a HRSG downstream of a gas turbine (firing natural gas or No. 2 oil) is 5 to 7 fins per inch (200 to 280 fins per meter) with a foreseeable development potential of up to 340 fins per meter. If the gas turbine uses heavier fuels, the fin density is reduced to 3 to 4 fins per inch (120 to 160 fins per meter) to better handle deposits.

The optimum HRSG must fulfill the following, sometimes contradictory, conditions:

- The rate of heat recovery must be high (high efficiency)
- Pressure drop on the exhaust side of the gas turbine must be low to limit losses in power output and efficiency of the gas turbine
- The permissible pressure gradient during startup must be steep
- Low-temperature corrosion must be prevented

It is particularly difficult to meet the first two conditions at the same time. Because of the relatively low temperature, the heat transfer takes place mainly by means of convection. The differences in temperature between the exhaust and the water (or steam) must be small to obtain a good rate of heat recovery, which requires large surfaces. This means a large pressure drop in the exhaust unless the velocity of the gas is kept low, which would result in a lower heat transfer coefficient and, therefore, further increase the surfaces required. Using small-diameter finned tubes helps to solve this problem.

Another benefit of small-tube diameters is the small amount of water in the evaporator, resulting in a lower thermal constant, which favors quick load changes.

HRSGs being built today have low pinch points and small pressure drops on the exhaust side. Pinch point values of 6 to 15K (11 to 27°R) at pressure drops of 25 to 35 mbar (10" to 14" W.C.) are attainable even with a triple-pressure reheat boiler.

Low-temperature corrosion. When designing HRSGs, care must be taken to prevent or restrict low-temperature corrosion. To accomplish this, all surfaces that come into contact with the exhaust must be at a temperature above or slightly below the sulfuric acid dew point. When burning a sulfur-free fuel in the gas turbine, the limit is determined by the water dew point.

Suitable precautions enable operation of heat exchangers at temperatures below the acid or water dew points. This can be done by selecting appropriate materials, or by adding corrosion allowances to the design of affected tubes. Because there are only a few tubes in this temperature range, this strategy can be beneficial despite the low exergy of the heat gained.

Optimum design of a HRSG. Designing a HRSG involves optimizing between cost and benefit. The main cost driver is the heat exchange surface installed. The indicator generally used is the pinch point in the evaporator. The surface of the evaporator increases exponentially as the pinch point decreases, whereas the increase in steam generation is only linear. For that reason, the pinch point selected is a critical factor in determining the heat transfer surface. In installations where efficiency is highly valued, the pinch point is 8 to 15K (14 to 27°R); where efficiency is lower, the pinch point can be higher, 15 to 25K (27 to 45°R).

Operating experience. One challenge affecting the design of the HRSG is the quick startup capability of the gas turbine, especially for the cold and warm plant conditions. The rapid thermal expansion that occurs during a startup can be accommodated through suitable design measures such as suspension of the tube bundles, drum design, tube-to-header connections, and so on.

The main constraint for the loading gradient is often dictated by the drum. To enable as quick a start as possible, the walls of the drum should be as thin as possible, which can be achieved, provided the design live-steam pressure is sufficiently low. Modern gas turbines have higher exhaust temperatures than older models; therefore, higher live-steam pressures, especially with reheat steam cycles, are now more attractive and this results in longer startup times.

Once-through HRSGs eliminate the thick high-pressure drum and, therefore, afford the desired high thermal flexibility.

Another point of concern is the volumetric change in the evaporator during startup. The large differences in specific volume between steam and water at low and intermediate pressures cause large amounts of water to be expelled from the evaporator at the onset of the evaporation process. If the drum cannot accommodate most of this water, a large amount of water would be lost through the emergency drain of the drum during every startup, or an undesired emergency trip of the unit would be required to prevent water carryover into the steam system.

To improve part-load efficiency and behavior of the combined-cycle plant, the boiler is operated in sliding-pressure mode. That means the system is generally operated at a lower pressure when the steam turbine is not at full load. This can be accomplished by keeping the steam turbine control valves fully open. For example, in a system with two gas turbines and two HRSGs feeding a common steam turbine, half-load of the whole power station can be accomplished with only one of the gas turbines running at full load. In sliding-pressure mode, the live-steam pressure is at 50% of the pressure at full load. The steam volumes in the evaporator, superheater, and live-steam lines of the HRSG in operation are doubled.

During off-design conditions, economizers can start to generate steam, which can block tubes and reduce the performance of the HRSG. To keep this within limits, the economizer is dimensioned so that the feedwater at the outlet is slightly subcooled at full load. This difference between the saturation temperature and the water temperature at the economizer outlet is known as the approach temperature. Because it causes a reduction in the amount of steam generated, it should be kept as small as possible (typically 2 to 12K [4 to 22°R]). Proper routing of the economizer outlet tubes to the drum also prevents blockage if there is steaming in the economizer.

Another way of preventing steaming in the economizer is to install the feedwater control valve downstream of the economizer; the economizer is thus kept at a higher pressure and steaming is prevented.

HRSG with limited supplementary firing

Despite the fact that the majority of the HRSGs are unfired, there are occasional applications where a limited amount of supplementary firing is required.

The operating principles of a HRSG with limited supplementary firing are the same as those for the unfired boiler. There are various designs available for the firing system itself. Units that do not exceed a gas temperature of approximately 780°C (1436°F) downstream of the supplementary firing can be built with simple duct burners. This limit can be extended but requires modifications to the design of the HRSG.

Figure 7–10 shows a HRSG with supplementary firing. This system is particularly well suited to combusting natural gas, which attains uniform temperature distribution downstream of the burners. Radiation to the walls of the combustion chamber is relatively low. For that reason, the majority of the HRSGs of this type combust natural gas. There are systems available for oil, but the burner equipment is more complex and costly to install, operate and maintain.

The importance of supplementary firing in HRSGs used for power generation alone is diminishing. This is mainly caused by two facts:

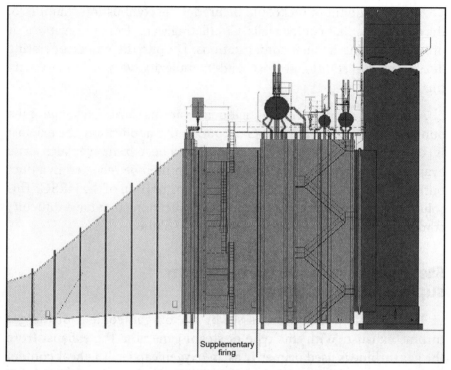

Figure 7–10 Supplementary fired heat recovery steam generator

- Modern gas turbines have exhaust temperatures closer to the maximum allowable HRSG inlet gas temperature, thereby reducing the effect of any supplementary firing.

- Supplementary firing downstream of a modern gas turbine results in an efficiency decrease for the combined cycle. With older gas turbine models, the efficiency was more or less constant.

Supplementary firing is most often applied in combined-cycle cogeneration plants where the amounts of process steam must be varied independently of the electric power generated. In this case, supplementary firing is used to control the amount of process steam generated. Additional applications include combusting gases, which are not suitable for firing in the gas turbine (e.g., pressure is too low or heat content unsuitable), and reaching a higher power output, but with the expensive penalty of a lower power generation efficiency.

A larger amount of fuel could be fired if the combustion chamber of the duct burner has cooled walls. Circulating water from the evaporator provides cooling in such configurations. Despite the higher operating flexibility, the additional cost and complexity very seldom justify this solution.

Arranging supplementary firing in various heat exchange tube bundles can also increase HRSG output. In this application, the exhaust is cooled downstream of the gas turbine by a heat exchanger such as an evaporator or superheater. The exhaust can then be reheated in a duct burner before passing through the remaining portion of the HRSG. This solution is rare and reserved for niche applications that have difficulty competing with an unfired HRSG or other solutions.

Steam generator with maximum supplementary firing

The maximum firing rate is set by the oxygen content of the gas turbine exhaust. With this type of steam generator, the exhaust from the gas turbine is used primarily as an oxygen carrier. The heat content of the gas turbine exhaust is low compared with the heat input of firing in the boiler. It is, therefore, no longer correct to speak of a HRSG.

The design of a steam generator of this type is practically identical to that of a conventional boiler with a furnace, except that there is no regenerative air preheater. The gas turbine exhaust has a temperature of 450 to 650°C (842 to 1202°F), rendering a regenerative heater unnecessary. To cool the exhaust to a sufficiently low temperature downstream of the steam generator, an additional economizer is provided, which takes over a portion of the feedwater preheating from the regenerative preheating of the steam turbine. The best arrangement splits the feedwater between the economizer and the high-pressure feedheaters. When the fuel is gas, an additional low-pressure partial flow economizer improves efficiency. The fuel combusted in the boiler may be oil, gas, or pulverized coal.

This application can be used to increase the output from an existing conventional steam turbine plant using a gas turbine and its exhaust energy. Due to the high integration of both cycles (e.g., the burners of the existing boiler must be retrofitted to make them suitable for the exhaust application), the complexity and the number of interfaces are considerable (see chapter 6).

Steam Turbine Technology

The most important requirements for a modern combined-cycle steam turbine are:

- High efficiency
- Short startup times
- Short installation times
- Floor-mounted configuration

In the past, combined-cycle steam turbines were applications of industrial steam turbines or derivatives from conventional steam turbine plants. The main differences between conventional steam turbines and combined-cycle steam turbines are:

- Higher power output
- Higher live-steam temperatures and pressures
- Higher number of extractions for feedwater heating

Today new combined cycles frequently are equipped with reheat steam turbines.

The main differences between combined-cycle steam turbines and conventional steam turbines are:

- Fewer or even no bleed-points as opposed to 6 to 8 for feed heating
- Floor-mounted configuration
- Shorter startup times
- Lower power output
- Lower live-steam pressures 100 to 170 bar (1430 to 2460 psig) as opposed to 160 to 300 bar (2310 to 4340 psig)

With the different requirement profile and the large volume of the combined-cycle plant orders, an optimized design for this application is justified.

Characteristics of combined-cycle steam turbines

Steam turbines used for combined-cycle installations are simple machines. In the past they used relatively low live-steam parameters; with increasing gas turbine exhaust temperatures, optimal steam pressures also increased. Live-steam temperatures have now reached the level of those at conventional steam power plants.

Combined-cycle plants frequently generate steam at more than one pressure level. Due to multiple inlets, the steam mass flow in the steam turbine increases from the inlet towards the exhaust (the bleed for the partial feedwater preheating involves only a small reduction). In a conventional steam turbine, the steam-mass flow is reduced to roughly 60% of the inlet flow at the exhaust.

Short startup times are of particular importance because combined-cycle plants are often used as medium-load units with daily or weekly startups and shutdowns.

Table 7–4 illustrates how in recent years, the combined-cycle plant has profoundly changed gas turbine technology, resulting in a major impact on steam turbine requirements:

Table 7–4 Change of boundary conditions for steam turbines in combined-cycle plants

	Traditional	Modern
Gas turbine exhaust temperature	450–550°C (842–1,022°F)	550–650°C (1,022–1,202°F)
Live steam temperatures	420–520°C (788–968°F)	520–600°C (968–1,112°F)
Number of steam pressure levels	1 or 2	2 or 3
Live steam pressure	30–100 bar (420–1,430 psig)	100–170 bar (1,430–2,460 psig)
Reheat steam cycle	no	yes
Number of gas turbines per steam turbine	1 to 5	1 or 2

Multishaft and Single-Shaft Plants

In a multishaft combined-cycle plant, there are generally several gas turbines with HRSGs generating steam for a single-steam turbine. The steam and gas turbines use separate shafts, generators, step-up transformers, and so on. By combining the steam production of all the HRSGs, a larger steam volume enters the steam turbine, which generally raises the steam turbine efficiency.

Modern gas turbines achieve higher output with high exhaust temperatures. With the large gas turbines on the market, one steam turbine per gas turbine or one steam turbine for two gas turbines is common.

If one steam turbine per gas turbine is installed, the single-shaft application is the most common solution—gas turbine and steam turbine driving the same generator.

A plant with two gas turbines can be built either in a two-gas-turbine-on-one-steam-turbine configuration (multishaft) or as a plant with two gas turbines, each in a single-shaft configuration.

There are two concepts for single-shaft plants. In the first, the generator is situated between the gas turbine and the steam turbine, each turbine driving one end of the generator, the steam turbine is coupled to the generator rigidly or with a clutch. In the second, the generator is at one end of the line of shafting, driven by both turbines from the same end. The steam turbine, therefore, is rigidly coupled to the gas turbine on one end and to the generator on the other. Figure 7–11 illustrates various single-shaft arrangements.

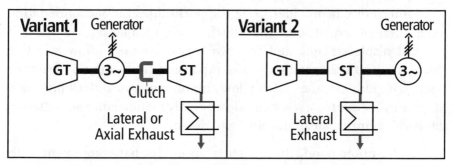

Figure 7–11 Different single-shaft combined-cycle configurations

The type of turbine-generator configuration (multishaft or single-shaft, with or without clutch) has a significant impact on components used in the respective combined cycle.

Single-shaft configurations with the generator situated between the two turbines also require a connecting flange on the generator exciter end that is designed for handling the steam turbine torque.

Steam turbines arranged between gas turbine and generator in single-shaft units also need a connecting flange on the high-pressure turbine end and the rotor must be able to transfer the torque of the gas turbine. Steam turbines located at the end of single-shaft power trains are mostly connected to the generator on the high-pressure turbine end and have an axial steam outlet at the end.

Clutch

A combined-cycle single-shaft configuration with the generator between the two turbines enables installation of a clutch between steam turbine and generator. The clutch is normally of the synchronous self-shifting type. This means the clutch engages in that moment when the steam turbine speed tries to overrun the rigidly coupled gas-turbine generator and disengages if the torque transmitted from the steam turbine to the generator becomes zero.

The clutch allows startup and operation of gas turbine without driving the steam turbine. This results in a lower starting power requirement and eliminates certain safety measures for the steam turbine (e.g., cooling steam or sealing steam).

Furthermore, it provides design opportunities for accommodating axial thermal expansion. The clutch itself compensates a portion of axial displacements, and the two thrust bearings allow selective distribution of the remaining axial expansion (reducing tip clearance losses). In addition, the clutch allows more operational flexibility such as gas turbine simple-cycle operation or early maintenance activities on gas turbine during steam turbine cool down.

On the other hand, the clutch is an additional component with a potential impact on availability. Additionally, the generator located at the end of the line of shafting has advantages during generator overhaul and the rare necessity of rotor removal. However, single-shaft units without a clutch definitely need auxiliary steam supply to cool the steam turbine during startup. This is not necessary in units with a clutch.

Figure 7–12 shows a steam turbine for a single-shaft unit in a triple-pressure reheat plant. The steam turbine output is 142 MW with live-steam parameters of 170 bar (2460 psig) and 565°C (1049°F).

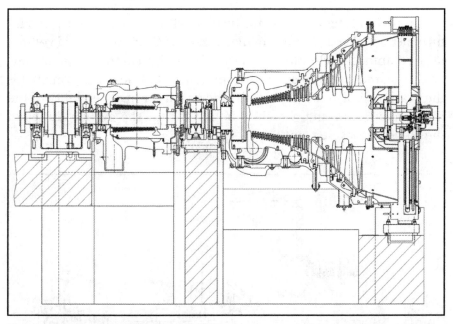

Figure 7–12 Cross section of a 142 MW reheat steam turbine with a separate HP turbine and a combined IP/LP turbine with axial exhaust

Live-steam pressure

Early gas turbines had lower power outputs and with their lower exhaust temperatures, the optimum pressure for the steam cycle was low. This was beneficial for the steam turbine for two reasons: low pressures yield larger steam volumes that can be more efficiently expanded in the steam turbine, and with a non-reheat cycle, a low live-steam pressure limits the moisture content in the steam turbine exhaust and, therefore, erosion.

Modern gas turbines with their high exhaust temperatures result in combined cycles with high live-steam pressures in the steam cycle.

A geared high-pressure steam turbine enables high-efficiency expansion with low live-steam volumes by using a gearbox to increase the speed of the high-pressure turbine, increasing blade length, reducing secondary blade losses, and resulting in an improved overall efficiency.

Figure 7–13 depicts a steam turbine with a geared high-pressure turbine that takes advantage of high steam pressure/low steam volume in a combined-cycle plant. It also shows the compact dimensions of the geared high-pressure turbine. The additional investment for the gearbox is also justified by the increased thermal flexibility of the high-pressure turbine, which is important for the cycling duties of the combined-cycle plant.

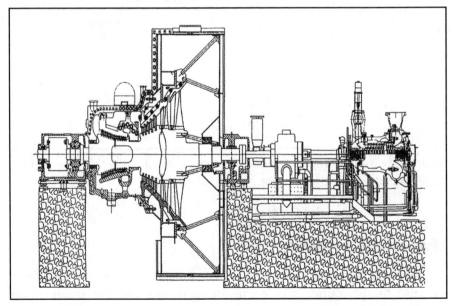

Figure 7–13 Cross section of a two-casing steam turbine with geared HP turbine

Generators

The majority of gas turbines and steam turbines are directly coupled to two-pole generators. For units with ratings below 40 MW, four-pole generators that run at half speed are more economical. It is advantageous that turbines in this output range usually already have a gearbox, therefore, using a four-pole generator only requires adaptation of the gearbox reduction ratio.

Four types of generators are used in combined-cycle plants:

- Air-cooled generators with open-circuit air cooling
- Air-cooled generators with a closed-circuit air (totally enclosed water to air-cooled, TEWAC)
- Hydrogen-cooled generators
- Water-cooled generators in single-shaft plants

Generators with *open-circuit air-cooling* are low in cost and have no need for additional cooling; however, problems with fouling, corrosive atmospheres, and noise can arise. Generators with *closed-circuit air-cooling* are built for capacities up to 480 MVA. These machines are reasonable in cost and provide excellent reliability. The full-load efficiency of modern air-cooled generators is above 98%. *Hydrogen-* and *water-cooled generators* attain efficiencies of approximately 99% at full load, making their efficiency performance superior, particularly at part-load, to air-cooled machines. The power density of water-cooled generators is greater, but their physical dimensions are smaller. However, water-cooled machines require additional auxiliaries and monitoring equipment, are more complex in design, and, as a result, more expensive than air-cooled machines. In general, generators achieve high reliability; air-cooled machines offer somewhat higher reliability because of their simplicity.

Figure 7–14 shows a generator with closed-circuit air cooling. Water is used to cool the air. These machines are well suited for use in single-shaft combined-cycle plants where the gas turbine and the steam turbine drive the same generator. Nowadays, air-cooled generators of up to 480 MVA have been built and tested due to the overall benefits of air cooling. For higher outputs, hydrogen or water-cooling must be selected.

Figure 7–14 Cutaway drawing of an air-cooled generator for use in combined-cycle power plants

Electrical Equipment

A single-line diagram for a combined-cycle power plant with a single shaft is shown in figure 7–15. It is similar to a single-line diagram for other types of power stations. Key features in this diagram are:

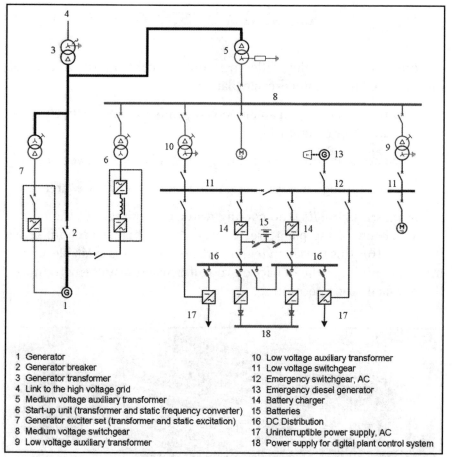

Figure 7–15 Single-line diagram

1. Generator
2. Generator breaker
3. Generator transformer
4. Link to the high voltage grid
5. Medium voltage auxiliary transformer
6. Start-up unit (transformer and static frequency converter)
7. Generator exciter set (transformer and static excitation)
8. Medium voltage switchgear
9. Low voltage auxiliary transformer
10. Low voltage auxiliary transformer
11. Low voltage switchgear
12. Emergency switchgear, AC
13. Emergency diesel generator
14. Battery charger
15. Batteries
16. DC Distribution
17. Uninterruptible power supply, AC
18. Power supply for digital plant control system

- One generator for the gas turbine and steam turbine (only one set of leads to the high voltage switchyard is needed)
- Auxiliaries fed from the generator output, which eliminates a separate feeder from the grid and allows the plant to run at full speed no load during a grid blackout, ready for reconnection
- Auxiliary transformers to feed the auxiliaries and to feed the static frequency converter to start the gas turbine using the generator as a motor
- Medium voltage bus for large motors
- Uninterruptible power supply for important loads

Cooling System

Three fundamentally different cooling configurations can be implemented in combined-cycle plants:

- Direct air cooling in an air cooled condenser (alternatively indirect air cooling)
- Evaporative cooling with a wet or hybrid cooling tower
- Once-through water cooling using river water or seawater

For *direct air cooling,* no cooling water is required, but the output and efficiency of the plant are reduced due to higher vacuum levels (figure 7–16). This is used in regions where no water is available, or as a means to minimize the impact on the infrastructure and environment, thereby facilitating permitting.

Figure 7–16 Typical arrangement of an air-cooled condenser

Alternatively, an indirect air cooling system could be used that employs dry cooling towers. Indirect cooling systems are more flexible in terms of arrangement, but usually also more expensive.

Evaporative cooling systems require water to replace evaporation and blowdown losses. This amount of makeup water depends on the exhaust steam of the steam turbine and amounts to approximately 0.3 kg/s (2400 lb/h) per MW installed plant capacity for a combined-cycle plant without supplementary firing.

An evaporative cooling system for combined-cycle plants is primarily of the mechanical draft type (wet-cell cooling tower as shown in figure 7–17). This is based on the fact that only half of the cooling capacity is necessary for combined-cycle plants compared to conventional steam power plants with an identical installed capacity. As the specific cooling demand is relatively small in combined-cycle plants, the required auxiliary power consumption for mechanical draft evaporative cooling systems is also low related to the gross plant capacity. On the other hand, the specific investment for natural draft cooling towers is very high, even for small cooling demands; therefore, mechanical draft cooling towers are most often the more economical option for combined-cycle plants (auxiliary power consumption versus investment).

Mechanical draft evaporative cooling systems for combined-cycle plants require approx. 0.5 to 0.7% of the gross plant capacity as auxiliary power consumption (cooling water pump and cooling fan power), whereas the demand for natural draft systems is less then 0.5%, which is comparable to direct air cooling with approx. 0.4 to 0.5% (for large ACC).

Figure 7–17 Typical arrangement of a wet-cell cooling tower

The once-through water-cooling variant has a water requirement that is on the order of 40 to 60 times greater than for evaporative cooling systems. After serving as heat sink, this water is returned to the water source (e.g., river, sea, or cooling pond) from which it was taken.

Normally, the decision on the type of cooling system used depends on the available supply of cooling water because the most economic cooling system is often a once-through water-cooled condenser. In a small number of cases, boundary conditions such as plume abatement or limitations prevailing at the site are determining factors when selecting the type of cooling system.

For example, a typical plume abatement concept for evaporative cooling systems is a hybrid cooling tower (figure 7–18). In hybrid cooling towers, 10 to 15% of the waste heat is rejected in a dry air cooling section, and the remaining waste heat in a conventional mechanical draft evaporative cooling system. The hot dry air downstream of the dry section reduces the humidity in the evaporative section and thus prevents formation of a visible plume.

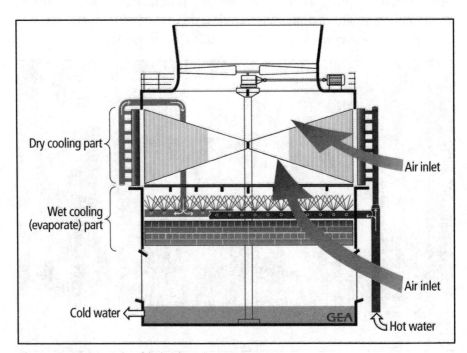

Figure 7–18 Principle of hybrid cooling tower

In a combined-cycle plant, however, the steam turbine generates only one-third of the installed electrical capacity (the gas turbine generates the other two-thirds), and requires only half the cooling capacity of a conventional steam power plant with an equal installed capacity. Hence the cooling system for combined-cycle plants has less impact on power output and efficiency when compared to the cooling system in conventional steam power plants.

Bypass Stack

Combined-cycle plants are sometimes equipped with exhaust bypass dampers and bypass stacks to permit for simple-cycle operation of the gas turbine(s). With modern gas turbines, this feature is becoming less and less important.

The reasons for this are:

- The design is expensive with gas turbine exhaust temperatures of 600°C and above.
- If the damper fails, the entire plant must be shut down.
- An exhaust damper has leakage losses with resultant losses in output and efficiency.
- The associated costs are significant.
- All these aspects become even more pronounced with advanced large gas turbines and the resulting larger dimensions, increased leakages, as well as higher exhaust temperatures.

An exhaust bypass stack (with or without damper) is a prerequisite for phased installation of a combined-cycle plant—one in which the gas turbine runs in simple-cycle operation before the steam cycle is connected. This allows the plant operator to get two-thirds of the final plant capacity on line sooner and thus create revenue at an earlier time. This can be economically attractive for plants with large capital payments. However, connection of the steam cycle requires interruption of normal gas turbine operation because the gas turbines are needed for commissioning of the steam cycle.

In deregulated markets, none of this is justified—the investment for the bypass stack, the expensive power generation during gas turbine simple-cycle operation, and the business interruption to connect the steam cycle—especially because a steam cycle can be installed in less than two years.

An exception might be feasible for locations with very low fuel costs and for countries with urgent need of new electricity supply. In such cases, phased construction could be economical also for advanced, large gas turbines. Nevertheless even for such cases a flue gas damper seems not to be recommendable due to previously mentioned inherent disadvantages. Use of simple blanking plates should be considered for changeover from simple-cycle operation to combined-cycle operation after completion of the water-steam cycle. For maintenance reasons, changeover to bypass operation will be possible within a reasonable time.

Instrumentation and Control

The complexity of a combined-cycle plant with its rapid interactions, especially during fast startup, requires a modern, advanced control system. Several power plant automation tasks such as turbine control for the gas turbine and steam turbine, boiler control, balance of plant (BOP), and integration of secondary equipment have to be performed in an optimized manner to maintain full control over the entire plant for all operating procedures involving transient processes and including startup and shutdown of the overall plant. Refer to chapter 8 for more information on this subject.

Other Components

In addition to the major equipment described in this chapter, the combined-cycle plant includes other equipment and systems. These include, for example:

- Fuel supply systems including a gas compressor, if required
- Steam turbine condenser
- Feedwater tank/deaerator
- Feedwater pumps
- Condensate pumps
- Piping and fittings
- Evacuation system
- Water treatment plant
- Compressed air supply
- Steam turbine bypass
- Civil engineering
- Heating, ventilation, and air conditioning
- Auxiliary boiler, if required

These are similar regardless of whether used in combined-cycle plants or in other types of power plants and, therefore, are not described here in detail.

8 Control and Automation

Power plants are normally operated to meet a demand dictated by the electrical grid, and where there is one, by the process steam system. The control systems discussed in this chapter are used to fulfill these requirements, to ensure correct control in transient modes such as startup and shutdown, and to ensure the safety of the plant under all operating conditions.

Control System

The control system is the brain of the power plant. Its tasks are the supervision, control, and protection of the plant. It enables safe and reliable operation.

The gas turbine is supplied with a standardized control system, providing a fully automated machine operation. The water/steam process is correspondingly automated to achieve a certain degree of uniform operation of the plant as whole, thereby reducing the risk of human error.

For this reason, the control and automation systems of a combined-cycle plant form a relatively complex system, even though the thermal process is fairly simple. Fully electronic control systems are applied in modern combined-cycle plants.

Main features of the control system for combined-cycle power plants are:

- Truly distributed architecture
- Complete range of functions for process control
- Communication capability due to several bus levels
- Compliance with standard communication protocols
- Openness for third-party applications
- On-line programmability with easy creation/editing of the programs

A hierarchic and decentralized structure for the open- and closed-loop control systems is best adapted to the logic of the whole process. It simplifies planning and raises the availability of the plant.

A highly automated control system encompasses the following hierarchic levels (see figure 8–1):

- **Drive level:** All individual drives are monitored and controlled. Drive protection and interlockings are provided by the control system. Signal exchange with higher hierarchical levels is also provided.

- **Functional group level:** The individual drives for one complete portion of the process are assembled into functional groups. The logical control circuit on this level encompasses interlocks, automatic switching, and preselection of drives. Examples of typical function groups are feedwater pumps, cooling water pumps, and lube oil system.

- **The machine level:** Includes the logical control circuits that link the function groups to each other. These include, for example, the fully automatic starting equipment for the gas turbine or the steam turbine.

- **The unit level:** This includes the controls that coordinate the operation of the whole combined-cycle power plant.

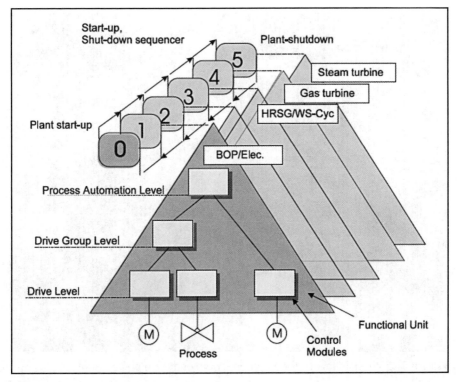

Figure 8–1 Hierarchic levels of automation

For a base-load power plant, a lower degree of automation could be selected because the frequent starting and stopping of an intermediate or peak load duty plant requires a higher degree of automation.

Process computers provide sequencing events, keep long term records and statistics, provide management information of the economy of the plant, optimize the heat rate and operation of the plant, and advise on the intervals between cleanings, inspections, and other maintenance work.

Figure 8–2 shows a standard layout for a modern combined-cycle power plant control room.

Figure 8–2 Standard layout for a modern combined cycle power plant control room

One type of control in a combined-cycle plant is closed loop control. This is when a controller receives actual measured data as an input, and uses it to correct a signal to a control device, with the aim of reaching the set point of a given parameter. The important closed control loops for a combined cycle fall into two main groups:

- The main plant load control loop
- The secondary control loops, which maintain the important process parameters, such as levels, temperatures, and pressures within permissible limits

A third group of closed control loops are component specific, such as minimum flow control for a pump or lube oil pressure control, and are intended primarily for the safe working of individual items of equipment. Because they do not directly affect the control of the cycle as a whole, they are not discussed here.

Applied logically, this leads to the hierarchical structure of the entire control system already mentioned.

Load Control and Frequency Response

The electrical output of a combined-cycle plant without supplementary firing is controlled by means of the gas turbine only. The steam turbine will always follow the gas turbine by generating power with whatever steam is available from the heat recovery steam generator (HRSG).

The gas turbine output is controlled by a combination of variable inlet guide vane (VIGV) control, and gas turbine inlet temperature (TIT) control. The TIT is controlled by a combination of the fuel flow admitted to the combustor and the VIGV setting. Modern gas turbines are equipped with up to three rows of VIGVs allowing a high gas turbine exhaust gas temperature down to approximately 40% GT load. Below that level, the turbine inlet temperature is further reduced because the airflow cannot be further reduced.

After a gas turbine load change, the steam turbine load will adjust automatically with a few minutes delay dependent on the response time of the HRSG. It is, however, sometimes suggested that independent load/frequency control of the steam turbine should be provided for sudden increases or decreases in load. Such a system would require the steam turbine to be operated with continuous throttle control, resulting in much poorer efficiencies at full and part loads and additional complications. Because the gas turbine generates approximately two-thirds of the total power output, a solution without control for the steam turbine power output is generally preferred. This is also supported by the fact that modern gas turbines react extremely quickly to frequency variations, and can usually compensate for the delay in the steam turbine response with falling frequencies.

Figure 8–3 illustrates the concept of closed loop load/frequency control without separate steam turbine control for a plant with two gas turbines and one steam turbine. An overall plant set point, Ps, is given to the overall plant load control system, KAR, which determines how the load should be distributed between the gas turbines. It receives a load/frequency signal LFC from each of the three generators, which is used to determine whether it is necessary to make any correction to the load. The load of the individual gas turbines is controlled by setting the position of the VIGVs and the TIT control, which varies the air and fuel flow to the gas turbine. Because the TIT cannot be directly measured, readings are taken of the turbine pressure ratio and exhaust gas temperature, from which the TIT is calculated.

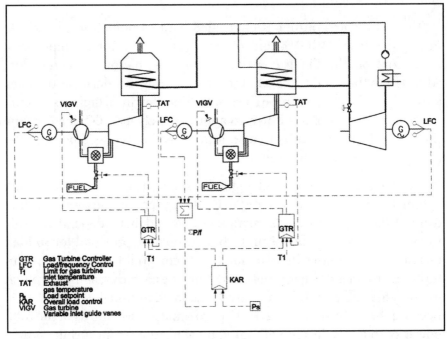

Figure 8–3 Principle diagram for a combined-cycle load control system

The entire steam cycle is operated in sliding pressure mode with fully open steam turbine valves down to approximately 50% live-steam pressure. This is the mode of operation best suited for high part-load efficiencies. More sophisticated control is not absolutely necessary because the power output of the plant can be adjusted by changing the set points of the individual gas turbine controls.

If supplementary firing is provided, it may be beneficial to provide independent load control for the steam turbine. The steam process then operates in a manner similar to that of a conventional steam plant, where the amount of steam generated is varied to fit demand by adjusting the supplementary firing fuel input.

Up to this point, no distinction has been made between load and frequency control. In principle, the remarks remain valid for both. However, another very important aspect of the load/frequency control is the capability of a plant to react to rapid fluctuations in frequency that may occur in the electrical grid. This is known as frequency response, and must normally be done in a matter of seconds, whereas loading takes place over several minutes.

To sustain stable operation of a plant, a grid frequency dead band of typically +/− 0.1 Hz is introduced within which the plant will not respond. Outside this dead band a droop setting is followed. The standard gas turbine droop setting is 5%, which means that a grid frequency drop of 5% would cause a 100% load increase. The droop characteristic setting is defined during the planning phase, and is typically in the range of 3 to 8%.

For plant configurations without steam turbine load control, the steam turbine will not be able to support falling frequencies within the 10 to 15 seconds normally required by grid codes, so the total response will have to come from the gas turbine. However, for increasing frequencies both the gas turbine and steam turbine will be able to support the grid. In such a mode the steam turbine valves will just close, allowing less steam to expand through the steam turbine.

Figure 8–4 shows a typical droop characteristic for a combined-cycle power plant with a 5% droop setting. If the frequency drops by 1%, the gas turbine load will initially jump 20% and the steam turbine load 0%. This result corresponds to a combined-cycle response of approximately 13%. On the other hand if the frequency increases by 1% the combined-cycle response of 20% is achieved by load changes of 20% in both the gas turbine and steam turbine, illustrated in the diagram by equivalent gas turbine load change of 27%.

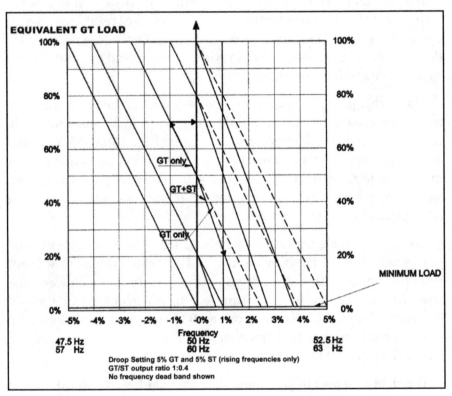

Figure 8–4 Typical combined-cycle droop characteristic of a GT load controller

To perform a plant load jump while the frequency is falling, it is essential that the gas turbine is operating below the maximum output level. Any operation, even in frequency support mode, above this level maximum is not possible. For frequency support gas turbines are typically operated between 50 and 95% load.

In conclusion, it can be said that combined-cycle plants are very well suited to rapid load changes. Gas turbines react extremely quickly because their time constant is very low. As soon as the fuel valve opens, more added power becomes available on the shaft. Gas turbine load jumps of up to 35% are possible, but they are not recommended because they are detrimental to the life expectancy of the turbine blading.

Secondary Closed Control Loops

Figure 8–5 shows the essential closed control loops that are required to maintain safe operating conditions in a dual-pressure combined cycle. The diagram shows the input signals into each controller and the command signals to the control valves. These control loops, which are typical for all combined-cycle concepts, are described as follows.

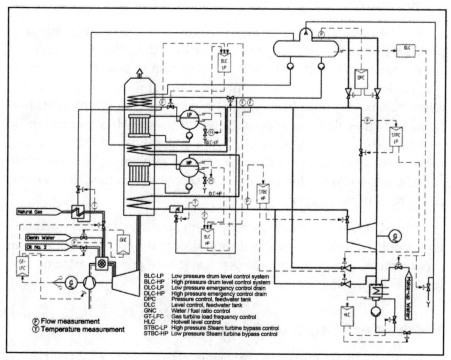

Figure 8–5 Closed control loops in a combined-cycle plant

Drum level control, BLC-HP, and BLC-LP

This is normally a three-element control system, as shown in figure 8–5 which forms one signal from the feedwater, the live-steam flows, and the drum level. This signal is then used to position the feedwater control valve.

Live-steam temperature control, STC

The live-steam temperature control loop in drum-type HRSGs actually limits rather than controls the live-steam temperature. Its purpose is to reduce the temperature peaks during off-design operating conditions, such as hot ambient temperatures, part load, and peak load. For that reason, attemperation often takes place after the superheater and not between two portions of the superheater as in a conventional steam generator. Normally, high-pressure feedwater is injected into the live steam to cool it down to the required temperature.

No real temperature control extending over a broad load range is possible in purely HRSG operation because the turbine exhaust gas temperatures drop off with cold ambient temperatures and extreme part loads of the gas turbine. In plants with supplementary firing, the relationships are more like those of a conventional steam generator. Because higher flue gas temperatures within the HRSG are possible in these cases, it is important that the temperatures of the steam and the superheater tubes be maintained within safe limits. To do this, the superheater could be divided into two sections with the attemporator installed in between them.

If the HRSG is of the once-through type, the live-steam temperature is generally controlled by the feedwater flow to the HRSG. With such an arrangement, however, attemperation is still required at extreme part load conditions and during startup.

Feedwater temperature, DPC

To avoid low temperature corrosion in the cold part of the HRSG, the feed water temperature should not, even in the lower load range, drop significantly below the acid or water dew point. On the other hand, this temperature must be as low as possible to ensure good utilization of the heat available, as discussed in chapter 5 "Combined-cycle Concepts". It is, therefore, recommended that the feedwater temperature be held at a constant level corresponding approximately to the acid dew point.

For the control loop shown on the diagram in figure 8–5, deaerator heating is done primarily with a steam turbine extraction, but if the pressure in this extraction is not sufficient at any load point then pegging steam from a live-steam source (LP in figure 8–5) is used. In plants with a low-pressure preheater loop in the HRSG, the opposite problem could occur during part load operation of the gas turbines, where more low-pressure steam could be generated than required. This excess energy must be dissipated either by increasing the pressure in the feedwater tank/deaerator, which reduces the amount of steam generated, or by directing the excess low-pressure steam to the condenser, which again slightly increases the vacuum and, therefore, marginally lowers the steam turbine output.

For plants without a feedwater tank, the minimum feedwater temperature is normally controlled by hot water recirculation from the low-pressure economizer outlet into the HRSG feedwater inlet (see figure 5–57). To compensate the pressure losses in the economizer, a small booster pump is foreseen. The amount of water is determined by the desired feedwater temperature. For high sulfur fuels (normally backup fuels), the economizer is bypassed and the water enters directly into the low-pressure drum.

Live-steam pressure, STBC-HP and STBC-LP

Usually the steam turbine is in sliding pressure operation down to approximately 50%, so continuous control for the live-steam pressure is needless in this range. Below this the pressure is kept constant by closing the steam turbine valves. Control is, however, necessary for nonsteady-state conditions such as startup, shutdown, or malfunction.

How this control is accomplished depends on the plant equipment. The live-steam pressure is controlled by the steam turbine control valves and/or steam turbine bypass control valves.

A 100% steam bypass provides the following advantages:

- Flexible operation during startup, shutdown, turbine trip, or quick changes in load
- Shorter startup times

- Environmental acceptability (because no steam is vented to the atmosphere)
- No actuating of the HRSG safety valves in case of a steam turbine trip

Table 8–1 shows when main steam turbine valves and steam bypass valves are used for various modes of operation of a combined-cycle plant with an unfired HRSG.

Table 8–1 Operation of steam turbine control loops in a single shaft combined cycle plant

Type of operation	Steam by-pass	ST control valves
Start-up	+	+
Shut-down	+	+
Normal sliding pressure operation	–	–
Fixed pressure operation below 50%	–	+
Steam turbine switch off or trip	+	–
HRSG switch-off	+	–
Gas turbine trip	+	–
+ in operation, – not in operation		

Level in feedwater tank and condenser hotwell, DLC and HLC

These levels ensure that there is sufficient head for the feedwater and condensate pumps and the necessary water buffer in the cycle condenser. Hotwell level is controlled by adjusting the main condensate flow control valve, which will open for increasing and close for decreasing hotwell levels. If the level in the feedwater tank is too high, the drain valve in the main condensate line is opened to prevent the level rising further. If the feedwater tank level is too low, makeup water is admitted to the cycle, usually via the condenser, which in turn raises the hotwell level, and thereby the feedwater tank level.

Supplementary firing

If the plant has supplementary firing, this will usually be controlled independently to meet a required output or process steam demand by regulating the amount of fuel admitted to the burners. This is often used to control variations in process steam flows, which cannot be achieved by the combined-cycle GT load control alone. If demand falls, the supplementary firing load is normally reduced first before the gas turbine load is decreased.

Process energy

If there is a process steam extraction, it is usually controlled independently of the main process control loops as previously described. Sometimes the control may even be done externally to the combined cycle, in the plant that is receiving the process. Usually, however, the pressure must be regulated, either internally or externally to the steam turbine, or using a combination of both. Sometimes the temperature must be controlled with attemperation.

Power augmentation

The various possible solutions are described and explained in chapter 9. The control of this system is done manually from the control room, and is normally not part of the plant load controller.

9 Operating and Part-Load Behavior

The way in which a power plant responds to changes in its outside conditions (ambient conditions, power demand by the grid) is of great importance for the overall plant economy as well as for a safe and reliable operation.

In a liberalized open market, gas turbines as well as combined-cycle power plants have the unique potential to react quickly and with flexibility to changes in grid and market requirements, and can, therefore, be used as *trading tool*. It is thus important to have the following:

- A well-designed power plant with the required flexibility (e.g., short startup time, high-loading gradients, possibilities for frequency support, good part-load behavior, additional system for power augmentation, etc.)

- Good knowledge (tool) to forecast accurate potential load jumps, or calculate cost of electricity at certain part-load conditions.

Combined-Cycle Off-Design Behavior

As opposed to the plant design, in which all the components of the water/steam cycle are defined to meet certain design criteria, the performance of a plant under off-design conditions will depend on the behavior of these fixed components with changes in the operating environment.

It is important to have precise knowledge of both the steady state and the dynamic operating behavior of the plant. Theoretical calculations of the dynamic behavior are costly and difficult. For that reason, to predict this behavior, operating experience from other similar plants or estimates (sometimes a simulator) are frequently used in combination with steady-state plant calculations.

A more exact calculation of the behavior would certainly be advantageous, but it is usually omitted due to consideration of time and cost. However, the calculations of the *steady-state operating and part-load behavior* have to be done.

Calculation basis

The calculation of the steady-state operating and part-load behavior of the steam part in a combined-cycle plant differs significantly from that of a conventional steam plant. The difference is related mainly to the boiler and the operating mode of the plant. In a heat recovery steam generator (HRSG), the heat is transferred mainly by means of convection, whereas in a conventional boiler it takes place due to radiation.

The steam turbine of a combined-cycle power plant functions most economically using sliding pressure, as shown in figure 9–1. (Sliding temperature process; that is, runs uncontrolled.)

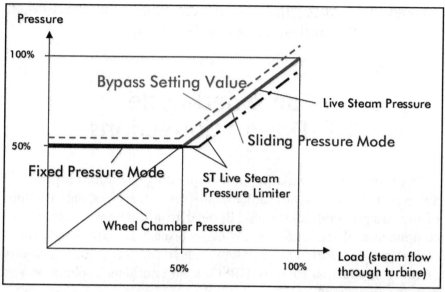

Figure 9–1 Sliding pressure diagram

Steam data are determined only by the exhaust flow and exhaust temperature of the gas turbine, and by the swallowing capacity of the steam turbine. In contrast, a conventional steam plant is generally operated at a fixed pressure; that is, live-steam pressure and temperature remain constant. Calculations are thus simplified because steam pressure and steam temperature are known in advance, and the steam turbine and the boiler can, therefore, be considered independent of each other. Calculations of the gas turbines are not a problem because gas turbines are standardized machines for which correction curves (or calculation tools) are available to account for changes in ambient conditions as well as for part-load operation.

Solution approach

Calculating the operating behavior of an installation, considering the geometry of this plant, would be very time-consuming. The process can, however, be simplified by referring all values to the thermodynamic data at the design point. If the design point is known, general equations (the Law of Cones, heat transfer law, etc.) can be used to reduce the calculation problem to a reasonable number of equations, without the necessity of considering the dimensions of the unit itself. A brief study of this calculation method can be found in appendix A.

Combined-Cycle Off-Design Corrections

The influence of a variation of the operating conditions is generally different from the behavior shown in chapter 5. Therefore, the variations of the design conditions generally lead to a different design of the components of the water/steam cycle (exceptions are standardized plants). In the case of off-design operation, the actual geometry of the components, such as HRSG, steam turbine, cooling system, remains unchanged.

Operating behavior will differ from plant to plant, depending on the actual design point. In the following paragraphs, some typical variations are shown. These parameters should normally be considered when correcting the combined-cycle performance from one set of operating conditions to another (e.g., for combined-cycle performance testing):

- Plant load
- Ambient air temperature
- Ambient air pressure
- Ambient relative humidity
- Cooling water temperature (only in case of fresh-water cooling)
- Frequency
- Power factor and voltage of the generators
- Process energy extraction (only for cogeneration application)
- Fuel type and quality

Effect of ambient air temperature

At the design point, ambient air temperature has a large influence on the power output of the gas turbine and the combined-cycle plant (see chapter 5). Operating behavior of the plant in off-design conditions with changes in ambient air temperature is very similar to the behavior for different design points.

In particular, efficiency increases slightly if the air temperature increases (under the assumption that the vacuum within the condenser remains constant, as shown in figure 5–6). However, this effect is marginal and mainly affects the pinch point in the HRSG and condenser. These changes in performance are so slight and hardly detectable on the curve compared to the design influence discussed in chapter 5.

In case of combined-cycle plants with direct air cooling (e.g., with air-cooled condenser) or a wet cooling system, the ambient temperature will also have an influence on the condenser pressure and, consequently, on the steam turbine output and combined-cycle efficiency. Such a correction curve is plant specific (indicative values can be taken from figure 5–11 and figure 9–2.)

Effect of ambient pressure

The main factor influencing the ambient pressure is the site elevation, which is purely a design issue, as described in chapter 5. For a given site the power plant may see daily weather variations, which are the only causes of change in the ambient pressure. These corrections are basically the same as described in chapter 5. The correction will only affect plant output, and the efficiency will remain practically constant.

Effect of ambient relative humidity

The effect of changes in the ambient relative humidity is also similar to the one described in chapter 5. An increase in relative humidity increases the enthalpy of the working media of the cycle as well as the density of the air into the gas turbine. Compared to the design point it enhances the energy to the HRSG, causing slightly higher energy transfer through the HRSG sections. This leads to a marginal increase in the pinch point of the HRSG to transfer this additional energy, resulting in a slight negative tendency for off-design calculations.

In the case of combined-cycle plants with a wet cooling system the relative humidity will also have a larger influence on the condenser cooling water inlet temperature and, consequently, on the condenser pressure, steam turbine output, and combined-cycle efficiency. Such a correction curve is plant specific (indicative values can be taken from figure 5–11 and figure 9–2.)

Cooling water temperature

The cooling water temperature has a major impact on the efficiency of the thermal cycle, directly affects the condenser pressure, and, consequently, affects the enthalpy drop in the steam turbine and the steam turbine output. At a given steam turbine exhaust area (selected at the design point) a change in the cooling water temperature affects with the corresponding condenser pressure the volume flow of the steam turbine exhaust, the steam velocity in the steam turbine exhaust part, and, consequently, the exhaust losses. Deviating from the design point the exhaust steam volume flow will be different as well as the exhaust losses. Generally exhaust losses are increased if the cooling temperature (condenser pressure) falls, so the benefit due to a better condenser vacuum is reduced. If the cooling water temperature is higher, the condenser pressure increases, thereby reducing the steam turbine output. The operating behavior is thus quite different from the one to arise if the size of the turbine was in all cases adapted to the temperature of the cooling water (identical starting point in the exhaust losses curve), as it is the case in chapter 5.

Figure 9–2 shows the effect of the condenser vacuum (steam turbine backpressure) on the relative efficiency of the combined-cycle plant for a typical direct water cooling (and wet cooling tower with reduced steam turbine exhaust area) application. It is a plant-specific curve, based on a given steam turbine with a fixed exhaust area (designed for a certain vacuum). The curve is, therefore, only generally applicable for plants with identical design conditions. To determine the condenser backpressure of different cooling media temperature, figure 5–11 is valid as long as the plant runs at full load.

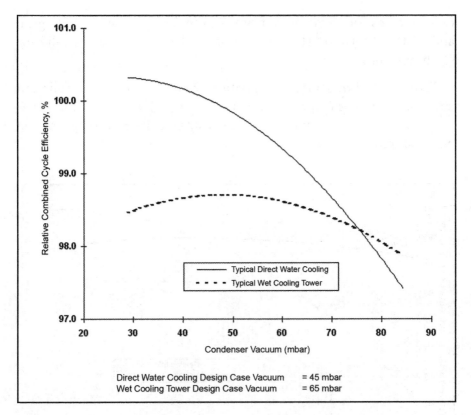

Figure 9–2 Effect of condenser vacuum on combined-cycle efficiency

For part-load conditions, less steam is produced in the HRSG, consequently reducing the amount of steam leaving the steam turbine exhaust into the condenser. If the cooling water flow is maintained for full as well as part loads, the vacuum is reduced even further for part loads due to the lower heating of the cooling media and the smaller pinch point in the condenser.

Electrical corrections

Frequency. The grid frequency has a major impact on plant behavior because it determines generator speed and, consequently, gas turbine speed. The gas turbine compressor speed defines the airflow entering the gas turbine, which is significant for the plant performance. Gas turbines are normally designed to operate at nominal firing temperatures for frequencies from 47.5 to 52.5 Hz for a 50 Hz grid, and 57 to 63 for a 60 Hz grid. The same design criteria are valid for steam turbines.

Figure 9–3 shows a typical variation of combined-cycle output and efficiency for frequency variations. The output decreases for lower frequencies and the efficiency stays within a narrow range of the nominal frequency point.

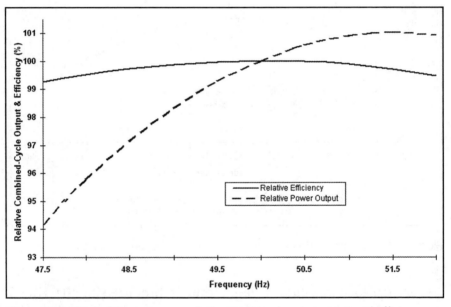

Figure 9–3 Effect of frequency on relative combined-cycle output and efficiency for full-load operation

Power factor. The plant power factors are dictated by the grid and influence the maximum generator capability as well as the generator efficiency, which in turn affects the output at the generator terminals. A normal power factor is in the range of 0.85 to 0.95. At the nominal point of the generator, a change in power factor from 0.8 to 1 would improve the generator efficiency by 0.3 to 0.4%. For lower loads of the generator the difference tends asymptotically towards zero.

Process energy. Plants with process heat extractions (cogeneration plants) are often highly customized, and universally valid off-design behavior cannot be given. The extracted process energy has a major effect on plant performance, making a plant-specific curve necessary (or calculation with an off-design heat-balance simulation program).

Fuel type and quality

The main off-design influence of the fuel on cycle performance occurs when a backup fuel is fired (e.g., oil instead of natural gas). The reason for this lies in the fuel composition and possibly the need for water or steam injection to meet local emission requirements.

Variations in the composition for the same type of fuel also influence the plant performance because a different fuel composition results in a different chemical composition after combustion.

Fuel components will also determine the lower heating value (LHV) of the fuel. If the LHV decreases, the fuel mass flow (heat input divided by LHV) increases to provide the same heat input to the gas turbine. This again results in increased flow through the turbine part of the gas turbine, which has a positive impact on output (provided the fuel has not first to be compressed in the plant).

If the chemical impact of the combustion products drags the performance in the opposite direction to the LHV influence, however, the total influence could be different. The normal tendency is, though, positive. It is, therefore, not possible to show an impact on a variety of fuels. These influences should be treated on a project specific basis, where the actual fuel compositions are known. A typical correction for gas fuel is shown in figure 9–4.

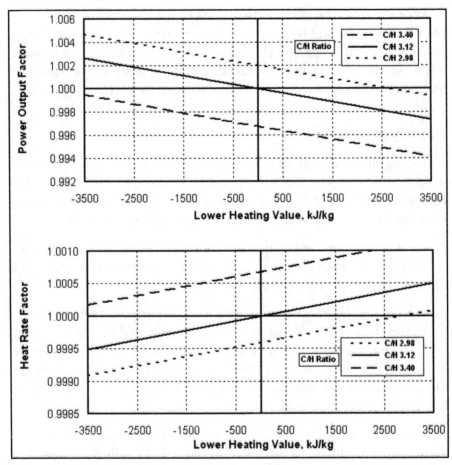

Figure 9–4 Effect of fuel composition and lower heating value on combined-cycle output and efficiency (base-load, gas operation with wet cooling tower)

Part-Load Behavior

In combined-cycle plants without supplementary firing, efficiency depends mainly on the gas turbine efficiency, the load of the gas turbine(s), exhaust gas temperature, and plant size. So far, all of the corrections discussed have been related to a plant performance with gas turbines running at full load (with nominal firing temperature).

Figure 9–5 shows the part-load efficiency of a combined-cycle plant and the associated gas turbine, each relative to the 100% load case. At higher loads the part-load efficiency is acceptable down to an approximately 50% load due to the following reasons.

The gas turbine used is equipped with one or even more rows of compressor variable inlet guide vanes, keeping the efficiency of the combined-cycle plant almost constant down to approx. 80 to 85% load. This is because a high exhaust gas temperature can be maintained as the air mass flow is reduced. Below that level, the inlet temperature (TIT) must be reduced, leading to an increasingly fast reduction of efficiencies.

Also, the steam turbine is calculated with sliding pressure mode down to about 50% load, also providing good utilization of the exhaust gas in this range. Below that point, the live-steam pressure is held constant by means of the steam turbine inlet valves, resulting in throttling losses and increasing stack losses.

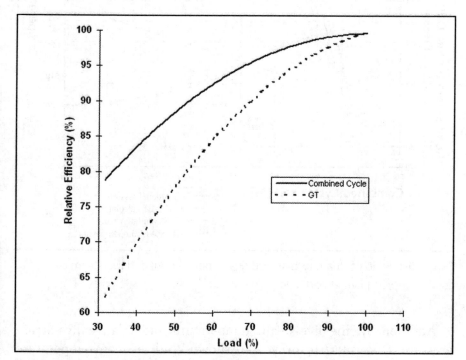

Figure 9–5 Part-load efficiency of gas turbine and combined cycle

Reheat gas turbines have a different (better) part-load behavior mainly at lower loads.

For full load operation the gas turbine accounts for two-thirds of the power output, and the steam turbine for one-third. Figure 9–6 shows how the ratio of steam turbine to gas turbine power output (PST/PGT) shifts towards more steam turbine output at part loads.

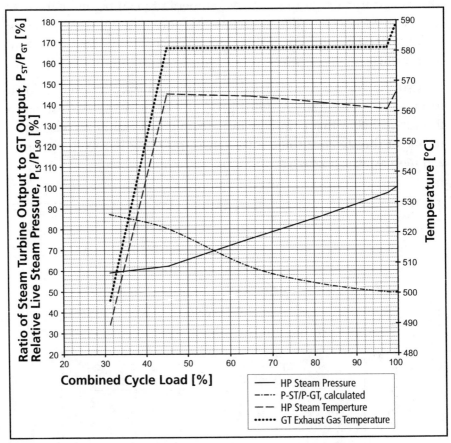

Figure 9–6 Ratio of steam turbine and gas turbine output and live-steam data of a combined-cycle plant at part load

Additionally, the live-steam temperature and relative live-steam pressure of the water/steam cycle are shown. The live-steam temperature is kept constant by means of attemperation at the HRSG superheater outlet. The reason for the flat temperature profile lies in the variable inlet guide vane control, which allows the gas turbine to operate at a lower flow with the nominal TIT. The live-steam pressure drops down to 50% of the full-load live-steam pressure and is then controlled by the steam turbine valves.

There are several site-specific possibilities for further part load efficiency improvements, such as:

- Several gas turbines in the plant configuration
- Air preheating for sites with cold ambient air temperatures

A combined-cycle plant with several gas turbines is operated differently at part load to keep the efficiency at highest possible level. For a plant with four gas turbines and one steam turbine, the load of the plant as a whole is reduced as follows:

- Down to 75%, there is a parallel reduction in load on all four gas turbines.
- At 75%, one gas turbine is shut down.
- Down to 50%, there is a parallel reduction in load on the three remaining gas turbines.
- At 50%, a second gas turbine is shut down.

With this mode of operation, the efficiency at 75%, 50%, and 25% load is slightly lower than at full load. If, however, four independent single-shaft combined-cycle blocks are selected, the part-load efficiencies are as shown in figure 9–7. In this case, the full-load efficiency would be achieved at the points 100, 75, 50, and 25% load because at these points the individual steam turbines are also running at full load.

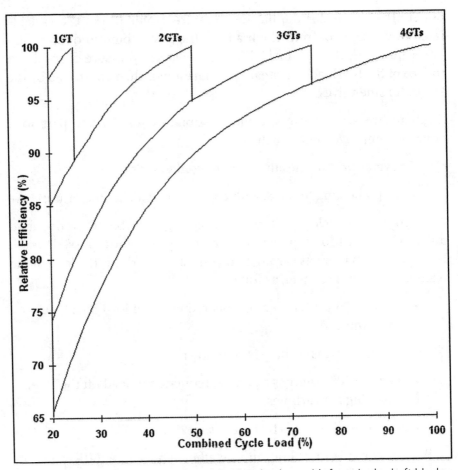

Figure 9–7 Part-load efficiency of combined-cycle plant with four single-shaft blocks

Combined-Cycle Testing Procedures

Plant performance guarantees are given when a power plant is sold. These guarantees will apply to a set of performance parameters and ambient conditions that cannot normally be recreated for the performance test. Therefore, to demonstrate that the guaranteed values have been met, plant performance must be measured under actual site conditions, and the results must be corrected with a series of corrections for the parameters described previously. These corrected values can then be compared with the guaranteed values.

In a combined-cycle plant, the gas turbine, HRSG, and steam turbine all interact with each other. If the contract for the plant goes to a single general contractor on a turnkey basis or to a single contractor for the power-island (including all main equipment such as gas turbine, boiler, steam turbine, generator, condenser, etc.), the power output and efficiency of the plant as a whole can be guaranteed. For combined-cycle plants, it is much easier and linked with fewer uncertainties to measure the values for overall plant performance instead of each major component individually. The amount of exhaust heat supplied to the HRSG (by the gas turbine in particular) cannot be measured accurately. When overall plant values are guaranteed, the fuel flow, lower heating value (with fuel samples), electrical output, and ambient conditions of the power plant must be measured. These are quantities that can be determined with relative accuracy. Nevertheless, a certain overall measurement uncertainty will remain; in case of a gas-fired combined-cycle power plant, about ±0.5% in overall power output and min ±0.8% in net efficiency.

Thereafter, a correction factor is determined for each parameter, quantifying its influence on actual performance because each parameter is not at design/guarantee value. These correction factors are multiplied to give values, which can be directly compared with the guarantees.

The power output of the gas turbine and the steam turbine are often corrected separately.

- For the gas turbine, the usual correction curves are used to take into account the effects produced by air temperature, air pressure, rotational speed, and so on.

- The power output measured for the steam turbine is corrected by using curves that show the indirect effects of air temperature, air pressure, and gas turbine speed on the steam process as well as the direct effect of the cooling water temperature (or air temperature and humidity).

To calculate these curves, it is best to use a computer model that simulates the steam process as a whole (as previously described in this chapter). Changes in ambient air data produce changes in the gas turbine exhaust data, and these affect the power output of the steam turbine.

The advantage of this procedure is that it can, with certain modifications, be used even if the gas turbine is put into operation at a somewhat earlier date than the steam turbine (phased construction). The method is, however, rather complicated and requires quite a few corrections to cover the interactions between the gas turbine(s) and steam turbine. Therefore, especially for single-shaft combined cycles and for plants without phased construction, there is a trend towards overall combined-cycle corrections.

For steam turbine plants and gas turbines and combined-cycle power plants the methods used for corrections are described in international standards such as ASME PTC 22 and 46 or ISO 2314.

Correction with a heat balance computer model: In cases of more complex combined-cycle plants (e.g., with district heating or process steam extractions), the correction procedures with correction curve for each and every parameter would be too complex. Therefore, it is easier and more accurate to make the correction from test results to guarantee condition (or vice versa) with a heat balance calculation program. Principles of correction are shown in figure 9–8. The correction from the measured condition back to guarantee condition (3 → 4) has an advantage in case of a weighted guarantee, resulting out of different individual guarantee points. Of course, the calculation model must be tested and jointly agreed by all parties prior to testing.

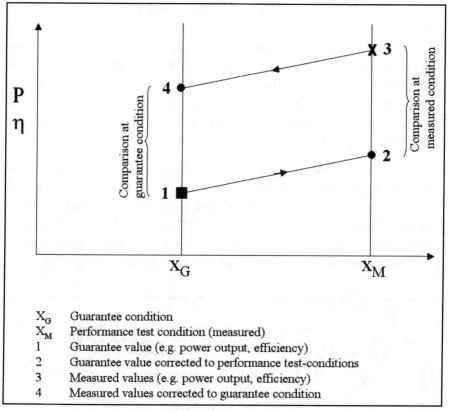

Figure 9–8 Performance guarantee comparison

Combined-Cycle Power Plant Degradation

Performance predictions or even performance guarantees are required for long-term economical evaluation to be given after a certain number of operating hours or years of operation. After commissioning at the time of the plant performance test, the plant is normally said in "new and clean condition". For operating hours beyond this point, the power plant performance value will slightly decrease as function of time due to degradation, which is mainly caused by the gas turbine and partially recuperated in the steam process. Typical values are given in chapter 7. Plant performance degrades from new and clean values because of many factors, including GT compressor fouling, increased leakages, and airfoil surface finish changes. Plant operation can also have a significant impact on performance. Off-line and on-line compressor washes, inlet filter replacements, frequency of start/stops, plant chemistry, readjustment of instruments, and controller setting values can be major factors.

Figure 9–9 indicates a typical degradation on power output and efficiency for a large combined-cycle plant with natural gas firing after GT compressor off-line washing and with a state-of-the-art maintenance and operation of the entire plant. The last one can be achieved by an operation and maintenance contract with the plant (GT) supplier, with clearly defined guarantees for degradation as well as for availability.

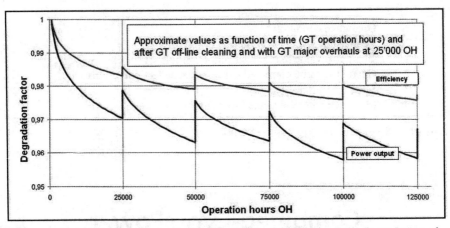

Figure 9–9 Expected non-recoverable combined-cycle power plant degradation of power output and efficiency with GT operating on clean fuels

As already explained in chapter 7, the degradation factor is a function of various elements. A few of them can be optimized (e.g., compressor washing, exchange of GT air intake filter, etc.) and kept as low as possible. To visualize the performance degradation over the plant operation time, a performance monitoring system in the control room can record specific values, and correct these data to reference conditions (see figure 9–10).

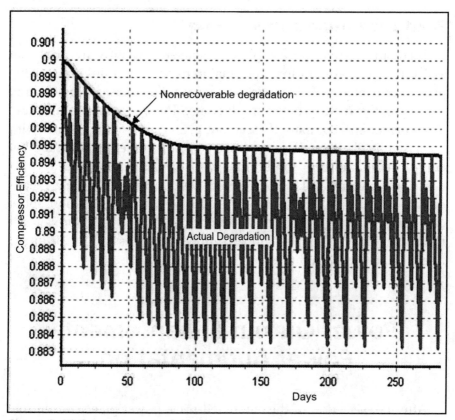

Figure 9–10 Gas turbine compressor efficiency as function of days operation (or operation hours)

Figure 9–10 clearly shows that the real average power output and efficiency at guaranteed condition over a defined interval (e.g., between PAC and first GT overhaul) is the base for an economical calculation, is not easy to forecast, and will, in most of the cases, not be guaranteed.

For a well-maintained and operated combined-cycle power plant the real figures are about 3 to 4% for the average power output, and about 2 to 3% below the values for efficiency according the curve shown in figure 9–9.

Plant performance monitoring

Such an on-line power plant management software tool can assist the local operation and maintenance team in maximizing plant performance (minimize permanent degradation losses) and consequently owners' operating income. The system's primary duty is to correct and compare the current plant performance values with the base line plant (mostly new and clean, or a predefined degradation curve) at guaranteed conditions, with a correction procedure as described under combined-cycle off-design corrections. By means of such a tool, the operator gets a clear and early indication about deviation to target performance, and he can react with improvement measures (e.g., off-line, on-line washing of compressor, GT air intake filter replacement, calibrate instruments etc.).

Power Output Enhancement-Power Augmentation

After the selection of the combined-cycle power plant concept has been done, other factors are addressed. These factors include:

- Is there a need for peak power production with premium paid for the resulting power?
- Does peak power demand occur on hot days (summer peaking) only?
- Is there a need to compensate the power reduction continuously during summer period?
- Is frequency support required?

All these questions must be checked individually. The goal of power augmentation basically is to increase the overall net power output of a combined-cycle power plant for a limited time (i.e., during midday peak with high ambient air temperature) with possibly some concessions regarding efficiency.

There are different possible solutions for power augmentation applications, such as:

- Supplementary firing in boiler
- Steam/water injection into the gas turbine
- GT peak load firing
- GT air inlet cooling

GT air inlet cooling is mainly for larger combined-cycle power plants a realistic economical option and use the effect, that with lowering of the air inlet temperature the power output will be increased (see correction curve power output as function of the air temperature in chapter 5).

The control of this system is usually done manually from the control room, but today, mainly in cases where these systems are used for load jumps, they are integrated in the overall plant load controller.

GT air inlet cooling

Because gas turbines are constant-volume-flow engines, they are very sensitive to changing ambient temperature and pressure. This reduction of plant performance at high ambient temperature is particularly unfortunate because of the retail price for electricity that usually rises rapidly with higher ambient temperatures. The resulting mismatch of demand and supply does have a negative impact on the profitability of combined-cycle power plants. Thus, a variety of options have been developed to particularly account for this problem, such as:

- Evaporative cooling
- Fogging
- Evaporation compressor cooling (over/high fogging)
- Chiller

An overview of the different air inlet cooling systems is given in table 9–1.

Table 9–1 Overview of GT air inlet cooling systems

	(Demin) water consumpt.	Power increase	Air humidity limitation	Net efficiency reduction	Spec. Investment cost
Evap. Cooling	X	X	XXX	X	X
Fogging	X	X	XXX	X	X
High-/over fogging	XX	XX	X	X	XX
Chiller	N.A.	XX	N.A.	XX	XXX

X low   XX mid   XXX high

Evaporative cooling. The aim of this system is to cool down the air entering the gas turbine by evaporation of water, which also increases the mass flow. The evaporative cooler increases humidity close to saturation. The best result in power increase can be achieved if the air inlet temperature is high and humidity is low.

An overview of cooling potential at different locations is shown in table 9–2.

Table 9–2 Max ambient temperature and corresponding cooling potentials at different locations

Location	Germany	Gulf-region (seashore)	Saudi-Arabia (mainland)	Tropical Region	South USA
Max air temperature of °C	38–40	46	55	36–38	44
Cooling Potential K	14–16	14–15	26–28	10–12	18

The amount of evaporated water depends on ambient air temperature and humidity and is, in case of ambient temperature of 40°C and 50% humidity with a typical evaporative cooler efficiency of 85%, about 0.4% of the gas turbine inlet air mass flow.

The corresponding relative power increase of a combined cycle is shown in figure 9–11.

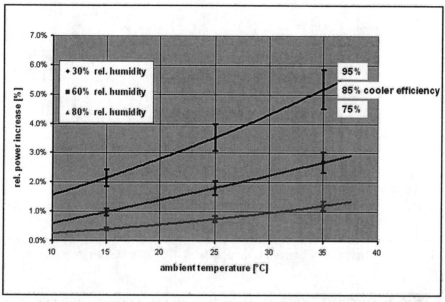

Figure 9–11 Relative power increase of a combined-cycle power plant as function of ambient temperature and humidity

For rough estimations the following could be applied: 1°C temperature decrease corresponds to a combine-cycle power increase of about +0.4 to 0.5%; the overall efficiency remains more or less the same.

An evaporative cooler solution does only make sense at locations with humidity below 70 to 80%. At air temperature levels lower than 10°C the cooling is usually not anymore in operation.

The system consists of an evaporative cooler (cellulose or fibreglass) and a droplet separator producing an additional pressure drop of about 1.5 to 3 mbar in the air intake system.

Fogging system. The principles are similar to the evaporative cooler. That means, the injected water (demineralized water) downstream of the air intake filter (see figure 9–12) cool down the inlet air with an improved cooling efficiency (up to 95% compared to about 85%). The injected water (small water droplets) evaporates and the air is cooled down close to saturation. In contrast to evaporative coolers, fogging systems have a negligible pressure drop and are ideal for retrofitting.

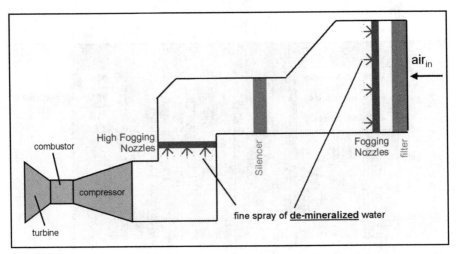

Figure 9–12 Typical arrangement of a fogging system in the GT air intake

High-fogging system. To further increase power augmentation, an additional nozzle rack is installed near the gas turbine compressor intake. These systems have different names such as high fogging, wet compression, over spray, or over-fogging system, and are usually in operation together with a fogging or evaporation cooling system.

The high-fogging system sprays small water droplets (<50μm) through nozzles into the air stream. These droplets evaporate mainly inside the compressor as the air is heated up during compression. Two effects mainly increase the power of the gas turbine:

- The air is intercooled in the compressor, which reduces the compression work and the compressor exit temperature.
- The mass flow through the turbine is increased.

The increase of power with fogging or evaporative cooling depends on ambient conditions. The high-fogging power increase is nearly independent of ambient humidity and temperature.

The total water mass flow capacity of a high-fogging system is gas turbine-specific and currently about 1.2% of the air intake mass flow at ISO condition, which results in a gas turbine power output increase of about 8 to 10%; combined cycle, about +5 to 7%. At high ambient temperature the total amount of water for air inlet cooling can go up to 2% of the air intake mass flow. The overall combined-cycle plant efficiency will be reduced by about −1.5% in case of a spray water amount of 2%.

To avoid erosion corrosion problems within the first part of the GT compressor, a well-designed and tested water spray system is mandatory to secure a uniform and fine water atomization into the GT air intake system. Additionally the first few rows in the GT compressor may need some modification for this kind of operation.

Chiller. Refrigerated air cooling is one of the most effective ways to increase the power of a gas turbine during high ambient temperatures. Chillers, unlike evaporative coolers, are not limited by the ambient temperature, and do not need additional water, which can be an important factor at dry area locations during the summer season.

The gas turbine inlet air can be cooled down to a fixed value (>10°C), which can be lower than the wet bulb temperature.

As shown in figure 9–13 the power consumption for the chiller unit increases drastically for cooling beyond the line when the air saturation line is reached (amount of work Δh_2 to cool down air from 23°C to 10°C is much higher than Δh_1 for cooling from 35°C down to 23°C).

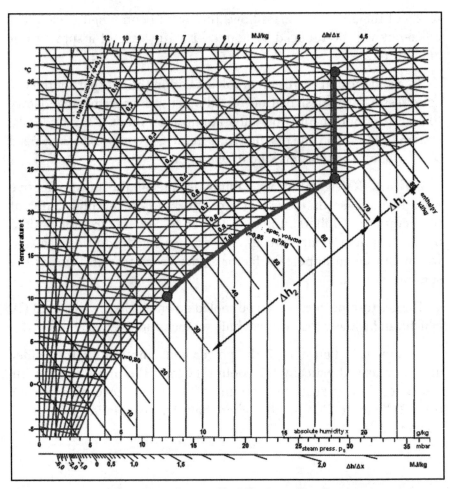

Figure 9–13 Air inlet cooling process with chiller in Mollier diagram

Most of the chilling energy in this case is used for the condensation of water and not for the temperature decrease. Therefore, an economical evaluation of the overall possible temperature decrease needs to be done by considering the additional cost as well as the additional power consumption for the chiller unit.

A typical diagram of a chiller system is shown in figure 9–14.

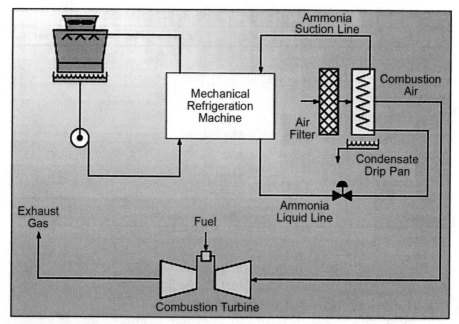

Figure 9–14 Typical diagram of a chiller system

The cooling of air can be provided either continuously or by using a thermal storage. Continuous cooling makes sense in cases where power enhancement is required for more than six to eight hours a day. A system with thermal storage is the better solution in all other cases. With such a thermal energy storage (TES) system, a cold reserve (in form of ice) can be built during the nonpeak hours and then be used for power increase during the few peak hours.

A typical example of a power improvement in a combined-cycle power plant with a chiller (without a TES design) is shown in table 9–3.

Table 9–3 Typical example of combined cycle power improvement with chiller

		Without chiller	With chiller
Design ambient temperature	°C	35	35
Design wet bulb temperature	°C	26	26
Design ambient RH	%	50	50
GT inlet air temperature	°C	35	10
Total net power output w/o chiller	MW	356	401
Parasitic power from chiller	MW		−13
Total net power output	MW	356	388
Net power increase	MW		32
% change	%		9.0%

The parasitic loads are in the range of about 20 to 30% of the power gain. In the example shown in table 9–3, the load is almost 30% because of the quite low cooling air temperature.

Summary

All systems, except the solution with chiller, need additional high-quality water (demineralized water) and are limited by the amount of moisture present in the air. Therefore, the application of these solutions is limited for sites with an average relative humidity of about 70 to 80%. The control of the different systems is done as follows:

- Evaporative cooler: No control of the evaporated water (only on-off possible)

- Fogging system: Air inlet temperature at GT compressor inlet can be controlled by the amount of injected water (with certain limits)

- Chiller system: Inlet air is cooled down to a fixed set point by a closed loop control (minimum temperature 10 to 13°C)

The high- or over-fogging (wet compression) solution has the largest potential for an increased power output by 10% (up to 20%). With this system special attention has to be given to a well-designed spray system (optimized droplet spectrum), including an adequate integration into the plant control system, as well as to the GT compressor (first 5 to 6 stages), to avoid extensive erosion/corrosion problems.

The specific costs of these additional systems are quite different. Table 9–4 gives a rough estimation, but as previously explained, mainly for the chiller system, various figures can heavily influence these costs.

Table 9–4 Specific investment costs for additional output of air inlet cooling system

		Spec. Investment Cost
Fogging System	US$/kW	25–50
Evaporative Cooing System	US$/kW	40–70
High Fogging System	US$/kW	120–160
Chilling System (w/o TES)	US$/kW	250–350

Startup and Shutdown of a Combined-Cycle Power Plant

Combined-cycle power plants are usually operated automatically. Therefore, during startup and shutdown it must be possible to activate equipment from the central control room. Whether the commands are to be issued to the individual drives, drive groups by the operating staff, from a higher level automatic starting program must be decided on a case-by-case basis. In base-load installations operating with few starts, full automation of the steam process is not always necessary.

The dynamic behavior of modern combined-cycle plants is characterized by the short startup time and quick load change capability. Above all, the gas turbine can be started and loaded quickly. Because its reaction time is short, it is capable of following quick changes and surges in load.

Modern combined-cycle plants in the 80 to 400 MW range can be started within the following times:

- Hot start after 8 hours standstill: 40 to 60 minutes
- Warm start after 48 hours standstill: 80 to 120 minutes
- Cold start after 120 hours standstill: 120 to 170 minutes

Larger figures correspond to 250 to 400 MW combined-cycle plants (one-to-one configuration)

Gas turbine startup is basically independent of its standstill time, allowing two-thirds of the combined-cycle power to be available within 30 to 40 minutes after activating the startup sequence. The startup time for the steam turbine depends on the time required to heat up parts of the machine (i.e., rotor) without exceeding thermal stress limits imposed by the material. Therefore, the stand-still time is very significant because it determines the temperature of these parts when the startup sequence is initiated. This temperature is the base for the allowable loading gradient of the steam turbine and the HRSG.

To increase the flexibility during plant startup a stress controller for the steam turbine (TSC) can allow different startup modes for the ST. It can be preselected by the operator, following grid requirements. Thermal stress and, as a result, lifetime consumption of the ST is cumulated, monitored, and indicate the remaining lifetime of the ST.

The operator has the choice between FAST, NORMAL and ECONOMIC loading modes. The FAST startup mode (maximum 800 ST starts) reduces the ST startup time, whereas the ECONOMIC mode (maximum 8,000 ST starts) saves ST lifetime consumption. The NORMAL loading is equal to standard cycling duties (maximum 2200 starts during ST life).

Starting procedure

To prevent explosion of any unburned hydrocarbons left in the system from earlier operation, it may be advisable—eventually mandatory by NFPA boiler code, especially when firing oil—to purge the HRSG before igniting the gas turbine. This is done by running the gas turbine at ignition speed (approximately 20 to 30% of nominal speed) with the generator as a motor, or with similar starting equipment, to blow air through the HRSG. The purge time (approximately five minutes) depends on the volume behind the gas turbine (e.g., bypass stack or HRSG), which has to be exchanged by up to a factor of five with clean air before gas turbine ignition can take place.

Plant startup begins after purging with the ignition of the gas turbine, running up to nominal speed, synchronized, and loading to the desired load.

An overview of expected values for unit startup times is given in table 9–5.

Table 9–5 Expected start-up times for a 400 MW combined cycle plant

Start-up condition	Expected values for unit start-up times*	ST operating mode
Cold start-up (plant shutdown for 120 h)	155 min.	FAST
Warm start-up (plant shutdown for 48h, over weekend)	105 min.	NORMAL
Hot start-up (plant shutdown for 8h, over night)	47 min. (31 min.)	ECONOMIC (fast)

*Plant start-up time begins with GT ignition and is completed with GT at IGV base load position and HP and IP/HRH bypass station closed.

The plant start-up time does not consider any pre-start preparations like filling of the water / steam cycle and HRSG, cold purging of the HRSG or waiting time for sufficient steam chemistry. Further it is assumed, that the plant was shut down without any disturbances and the steam turbine follows the regular cool down behavior (i.e. no forced cooling).

All systems of the plant have to run smoothly without upsets. The plant start-up time is achieved due to the high level of automation.

The operating regime is based on the optimized use of the turbine stress controller. NORMAL operation mode of the ST is determined by the design load regime as requested by the customer. The possible operating regime for FAST/ECONOMIC ST-operation mode used during start-up are resulting in higher respectively lower stress during ST start-up. With the selection of the start-up mode different start-up times can be achieved.

During gas turbine startup, depending on the actual startup condition (hot, warm, or cold), steam is generated more or less rapidly in the HRSG. In general, appropriate steam properties for a steam turbine startup are reached at approximately 50 to 60% gas turbine load. This means 40 to 60% of nominal pressure and a sufficient degree of superheat (e.g., around 50 K). Before starting the steam turbine, the gland steam system must be in operation and the condenser evacuated. Until the steam turbine takes over the available steam flow, the excess steam flows through the steam turbine bypass. If supplementary firing is installed, it should not be brought into operation until the gas turbine is at full load, respectively the steam turbine bypasses are closed and the steam turbine can accommodate the additional steam flow.

Figures 9–15, 9–16, and 9–17 show the three different startup sequences for a 250 to 400 MW class combined-cycle plant (in a one-GT-to-one-ST configuration), which can be brought to full load in as little as 40 to 50 minutes after eight hours standstill.

To ensure an early as possible GT load where the exhaust gas emissions are within acceptable limits and on the other side to produce with the exhaust gas energy in the HRSG sufficient steam (pressure and temperature) a shorter or longer holding point at around 40% GT load is set in all cases. A further loading of the GT up to full load or set target load is normally done at the point where the steam bypasses are closed and the steam turbine is able to admit all produced steam.

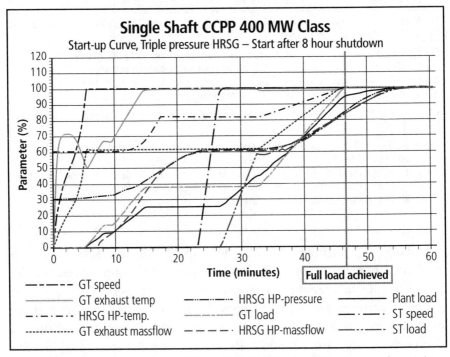

Figure 9–15 Startup curve for a 250 to 400 MW class combined cycle after eight hours standstill

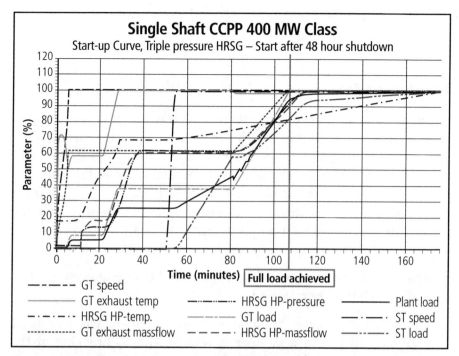

Figure 9–16 Startup curve for a 250 to 400 MW class combined cycle after 48 hours standstill

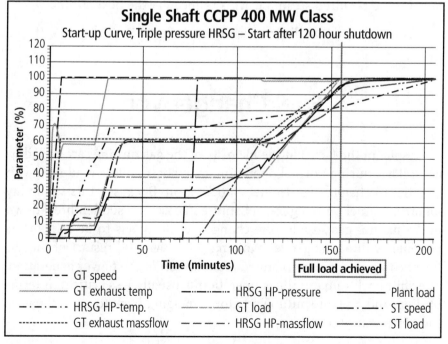

Figure 9–17 Startup curve for a 250 to 400 MW class combined cycle after 120 hours standstill

Plant shutdown

During unit shutdown, gas turbine and consequently steam turbine load is reduced according to the allowable transients to the inlet guide vane (IGV) point at nearly constant GT outlet temperature and approximately 60% of the exhaust gas/steam mass flow. At this point the GT load is kept constant and at full-steam temperature the shutdown program of the ST is started. The excess steam is dumped into the condenser via the respective bypass stations. At a preset load the ST trip is initiated to not cool down the ST unnecessarily.

After ST disconnection, GT load is reduced temperature controlled according to HRSG requirements.

To reduce the startup time after an overnight outage, closing of the HRSG outlet with a damper is recommended. Breaking of the condenser vacuum and shutting down of the gland (seal) steam system is recommended only in case of major plant outages. Thus the ingress of O_2 and CO_2 into the water/steam cycle is prevented, and the waiting time for the required steam purity is reduced at the next startup.

A typical shutdown diagram is shown in figure 9–18.

Fuel Changeover

Despite of the fact that the majority of the gas turbines are fueled by natural gas during normal operation, many of them are equipped with facilities to change over to fuel oil operation. Because of its suitability for storage, fuel oil is a good substitute in case of a sudden breakdown of the natural gas supply. The changeover delay has to be as short as possible, with a changeover avoiding shutdown and restart of the gas turbine to prevent an unplanned interruption of the power generation. The shift back to natural gas operation is usually less critical in terms of time, and a plant shutdown before changing back to natural gas is, therefore, acceptable.

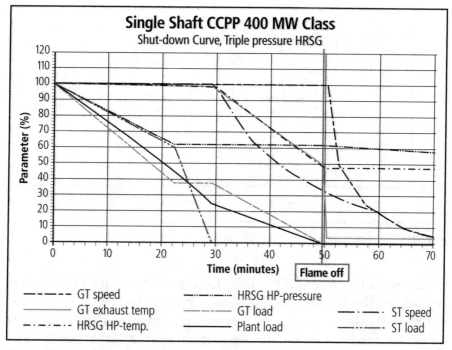

Figure 9–18 Combined-cycle shutdown curve

The overall time of the changeover is determined by the lead times of the fuel oil supply system, the gas turbine fuel oil system, and the changeover operation itself. The latter includes the following steps, and is shown in figure 9–19:

- Step 1: Load reduction from full-load operation (natural gas premix mode)
- Step 2: Change over to fuel oil diffusion mode
- Step 3: Switch over to fuel oil premix mode, followed by a load increase back to full load

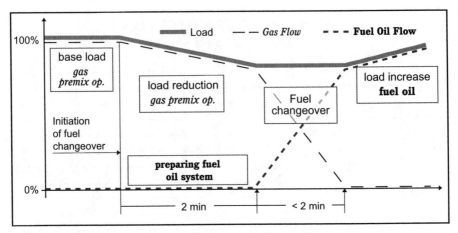

Figure 9–19 Switch over diagram from gas to oil operation

The transition from gas to oil fire in Step 2 can be done in the easiest way during the diffusion regime because no pilot burners are in service. Special attention should be paid to the maintenance of minimum flow rates of gas and oil to avoid flame pulsation, or even backfiring. Because of the enhanced NO_x emissions, the diffusion regime should be as short as possible.

The reverse sequence from oil to gas fuel operation needs water flushing of the closed oil burners to avoid clogging. The flow rate of the water is optimized between a maximum cleaning capability and a minimum impact on the stability of the gas flame.

10 Environmental Considerations

The impact any power plant has on its environs must be minimized. Legislation in different countries establishes rules and laws that have to be fulfilled. Quite often, emission limits are based on the best available emissions control technology. Exhaust emissions to the environment are mainly controlled in the gas turbine. Often, regions with less-stringent air emission requirements profit from the same combustion technology as areas with stringent requirements because the same hardware is used.

The following emissions from a power station directly affect the environment:

- Combustion products (exhaust and ash)
- Waste heat
- Waste water
- Noise
- Radioactivity and nuclear waste (nuclear power stations only)

Exhaust gases can include the following components: H_2O, N_2, O_2, NO, NO_2, CO_2, CO, C_nH_m (unburned hydrocarbons, UHC), SO_2, SO_3, dust, fly ash, heavy metals, and chlorides.

H_2O, N_2, and O_2 are harmless; the others, however, can negatively affect the environment. Concentration levels of these substances in the exhaust depend on the fuel composition and the type of installation.

However, the greater the efficiency of the installation the greater the dropoff in the proportion of emissions per unit of electrical energy produced.

Because most combined-cycle plants combust natural gas, they produce low exhaust emissions. Their high efficiency results in low emissions to the atmosphere per MWh of electrical power produced as well as a small amount of waste heat. The high excess air ratios typical of gas turbines enable practically complete combustion, which results in a very low level of unburned constituents such as CO or UHC in the exhaust. Because of the extremely low sulfur content of natural gas, SO_x (SO_2, SO_3) emissions are negligible. Particulate emissions from combined-cycle power plants are low due to the restrictions of solid matter input imposed by the gas turbine. Both fuel and combustion air have to be thoroughly cleaned to protect the gas turbine from corrosion and blockage of cooling-air passageways in air-cooled blades and in the combustion chamber. Therefore, a combined-cycle plant can be considered environmentally friendly and well suited for use in heavily populated areas.

For plants that combust natural gas, the most relevant emissions in the exhaust are NO_x (NO and NO_2). NO_x emissions generate nitric acid (HNO_3) in the atmosphere which, together with sulfuric and sulfurous acids (H_2SO_4, H_2SO_3), are factors that cause acid rain. CO_2 is created by burning fossil fuels and is considered responsible for global warming.

Reduction of NO_x Emissions

NO_x is produced in air in large quantities at very high temperature levels. Figure 10–1 shows NO_x concentrations in the air as a function of air temperature. The concentration shown is at equilibrium in air attained after an infinite time.

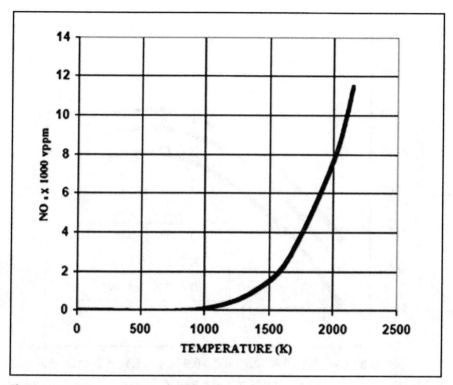

Figure 10–1 NO_x equilibrium as a function of air temperature

The situation in a gas turbine combustor is different because combustion of fuel in air takes place, and also because the high temperature residence time is fairly limited. The major factors affecting NO_x production in the combustor are:

- Fuel to air ratio of the combustion process
- Combustion pressure
- Air temperature in the combustion chamber
- Duration of the combustion process.

As seen in figure 10–1, NO_x is formed only when temperatures are high; this is the case in the flame in the combustor. The temperature of this flame depends on the fuel-to-air ratio of the flame and the air temperature in the combustion chamber, as shown in figure 10–2.

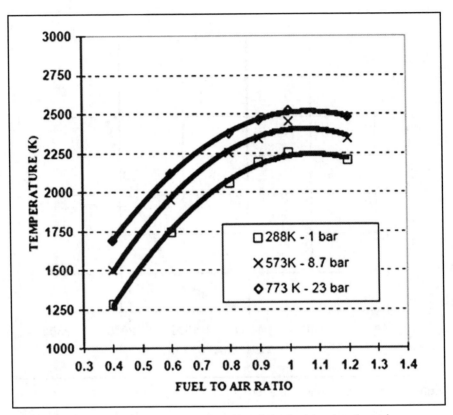

Figure 10–2 Flame temperature as a function of the fuel-to-air ratio and combustion air conditions

This temperature is highest in the case of stoichiometric combustion, (fuel-to-air ratio = 1). Figure 10–3 shows how concentrations of NO_x depend on the fuel-to-air ratio and the combustion air conditions.

It is evident that a peak is reached with a ratio of approximately 0.8. Above that level, the flame temperature is higher but there is less oxygen available to form NO_x because most of it is used by the combustion process. Below that level, NO_x decreases because of the abundance of excess air within the flame, thus lowering the flame temperature.

Very high fuel-to-air ratios are beneficial from the point of view of NO_x but are detrimental to combustion efficiency and cause the production of large amounts of CO and unburned hydrocarbons (UHC).

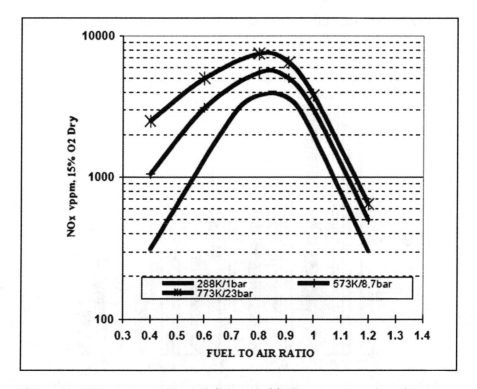

Figure 10–3 NO_x concentration as a function of fuel-to-air ratio and combustion air conditions

Conventional gas turbine combustors with diffusion burners were designed to operate with a fuel to air ratio of approximately 1 at full load, ensuring thorough, stable combustion over the entire load range. Obviously, NO_x emissions are high unless special precautions are taken. Nowadays burners operate at a lower fuel-to-air ratio that results in lower NO_x emissions.

The simplest way to reduce NO_x concentrations in these diffusion burners is to cool the flame by injecting water or steam into it. Figure 10–4 shows the reduction factors for NO_x emissions that can be attained as a function of the amount of water or steam injected. The amount of water or steam injected is indicated by the coefficient Ω (the ratio between the flows of water or steam and fuel). At a ratio of $\Omega = 1$, the typical reduction factor is approximately 6 with water and approximately 3 with steam. Water is more efficient than steam because evaporation takes place in the flame at low temperature, which enhances cooling efficiency.

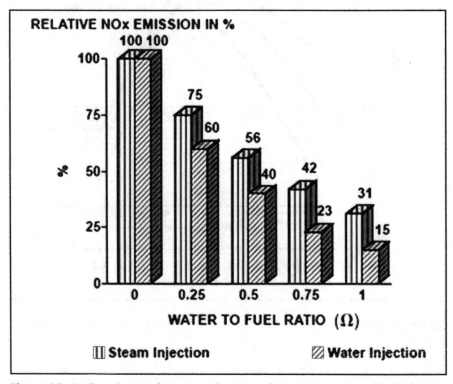

Figure 10–4 NO_x reduction factor as a function of the water or steam to fuel ratio in gas turbines with diffusion combustion

With this wet method, it is possible to attain NO_x levels as low as 40 ppm (15% O_2 dry) in the exhaust from gas-fired gas turbines or a combined cycle. Steam or water injection is a simple means to reduce NO_x emissions, but it does entail the following disadvantages:

- Large amounts of demineralized water are required.
- The efficiency of the combined-cycle plant is lower, particularly if water injection is implemented.

The fact that these methods can increase the plant output (especially with water injection) can be of interest in plants with a low number of operating hours per year. In general, however, this does not compensate for the loss in efficiency and the high water consumption (for further details see chapter 5, figure 5–9).

The disadvantages inherent to steam or water injection have motivated all gas turbine manufacturers to develop combustors which achieve low NO_x levels with dry combustion (i.e., without injecting steam or water).

The principle behind keeping NO_x levels low is to always dilute the fuel with as much combustion air as possible to maintain a low flame temperature, and to keep the residence time in the hot combustion zone short. With this dry low-NO_x method and firing of natural gas, NO_x levels below 25 ppm (15% O_2 dry) are achieved by modern gas turbines with high firing temperatures and efficiencies.

Theoretically, there are two ways that reduced NO_x emissions can be achieved:

- Oxygen-lean combustion
- Oxygen-rich combustion

Figure 10–3 shows the dependence of NO_x concentrations on the fuel-to-air ratio. Despite a high-flame temperature, only a small amount of NO_x can form in oxygen-lean combustion because there is hardly any oxygen available to produce NO_x. However, to attain complete combustion there must be a second, follow-up combustion stage here in which almost no NO_x is formed due to the lower temperature. This approach is being used in modern steam generators and is referred to as *staged combustion.*

For gas turbines with dry low-NO_x burners, a different—and more effective—procedure is applied because of the high overall excess air ratio (2.5 to 3.5). It is called *combustion with excess air* and is the same principle as the injection of water or steam, where a large amount of excess air effectively cools the flame. The procedure is subject to limits due to consideration of flame stability. With excess air ratios exceeding approximately two, combustion cannot take place and the flame is completely extinguished. This risk does not exist while the gas turbine is at full load because there is not enough air available for the burner to significantly exceed an excess air ratio of two. The remaining air is needed to cool the hot parts and the turbine blading. It is, however, a problem at part loads. For combustion to actually take place at the desired excess air ratio, the air and fuel must be mixed homogeneously with each other.

Figure 10–5 shows a typical low-NO_x burner. The air and natural gas are premixed in the burner and combusted downstream of the burner.

A vortex breakdown structure holds the flame in the free space. Characteristics of this burner include:

- No flashback due to the vortex breakdown structure
- Low fuel-to-air ratio (and, therefore, low NO_x)
- For liquid fuel, water injection is used to limit NO_x emissions

With this combustion technology there are modern gas turbines with high firing temperatures in operation with NO_x emissions clearly below 25 ppm NO_x at 15% O_2.

Further development of the low-NO_x burners is proceeding toward even lower emission levels. Due to investments in combustion research, lower NO_x levels are attainable and potential for further improvement is assumed.

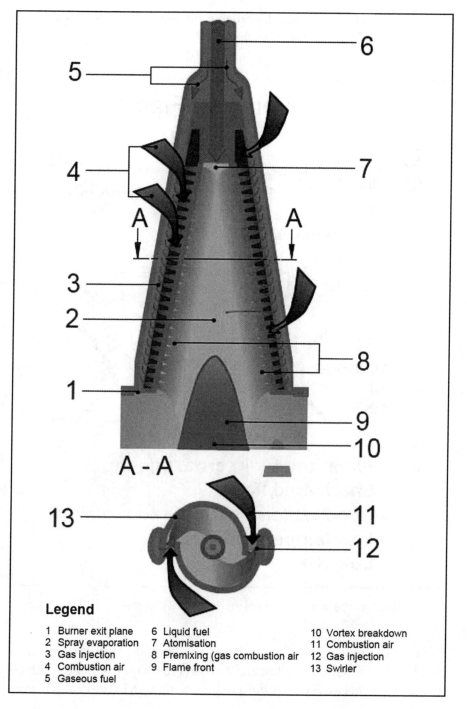

Figure 10–5 Cross section of a low NO$_x$ burner

Another technical concept available on the market, which follows the same principles, is shown in figure 10–6.

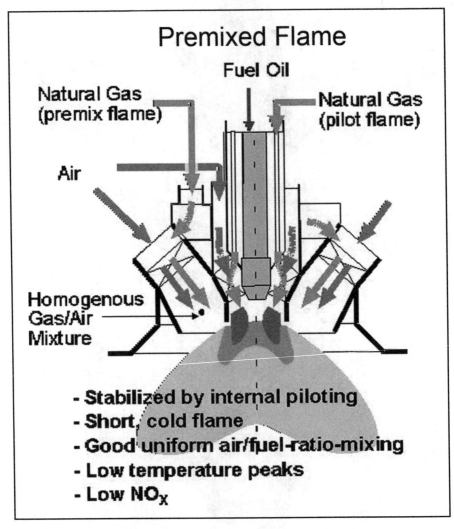

Figure 10–6 Cross section of the Siemens dry low NO_x burner

Local regulations in some parts of the United States and in Japan require NO_x emissions well below 25 ppm (15% O_2 dry). In these cases, it is generally necessary to install a reduction system in the heat recovery steam generator (HRSG). Known as selective catalytic reduction (SCR), these systems inject ammonia (NH_3) into the exhaust upstream of a catalyst and can thereby remove up to approximately 85% of the NO_x in the exhaust leaving the gas turbine. The chemical reactions involved are as follows:

$$4NO + 4NH_3 + O_2 = 4N_2 + 6H_2O \quad (10.1)$$

$$6NO_2 + 8NH_3 = 7N_2 + 12H_2O \quad (10.2)$$

Technically these are well-proven systems, but they entail the following disadvantages:

- Investment costs are high; the HRSG is 10 to 30% more expensive.

- The use of ammonia is necessary, with a fraction of the ammonia passing through the SCR (ammonia slip).

- Power output and efficiency of the power plant are reduced by approximately 0.3% due to the increased back-pressure of the gas turbine.

- When firing oil in the gas turbine, sulfur in the fuel reacts to form ammonium sulfate, which precipitates in the cold end of the HRSG and further increases back pressure to the gas turbine. Power output and efficiency of the plant is further reduced.

- The HRSG requires periodical cleaning, and the waste must be disposed of.

- The catalyst must be installed in the evaporator section of the HRSG because the reaction takes place only in a temperature window of 300 to 400°C (572 to 752°F).

- Replacement costs are high.

Figure 10–7 shows a typical SCR system installed in a HRSG. NO_x levels of less than 5 ppm (15% O_2 dry) can be achieved by implementing a SCR system in conjunction with a gas turbine equipped with dry low-NO_x burners.

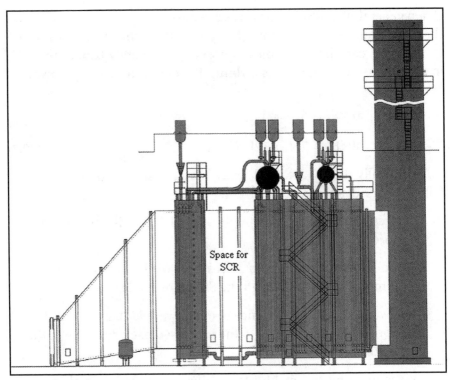

Figure 10–7 Heat recovery steam generator with selective catalytic reduction

SO_x Emissions

Concentrations of SO_2 and SO_3 produced depend only on the quality of the fuel. Approximately 70% of the gas turbines in operation use clean natural gas fuel; therefore, the SO_x emissions are negligible.

Liquid fuels contain between 0.05% sulfur in the case of sulfur-reduced No.2 oil (used only by few older gas turbines) and 2% sulfur for heavy or crude oil (used by approximately 4% of the gas turbine fleet). If the resulting SO_x emissions are not acceptable, the most economic way to reduce these emissions is to directly reduce the sulfur content in the fuel, either by chemically removing sulfur or by blending the fuel with a fuel with lower sulfur content.

CO_2 Emissions

Each fossil-fired power plant produces CO_2, which is considered responsible for global warming. However, a modern combined-cycle plant combusting natural gas produces approximately 40% of the CO_2 per MWh of electricity of a conventional coal-fired power plant. There are two reasons for this:

- The higher efficiency of the combined-cycle plant
- Use of natural gas fuel, which is mainly methane (CH_4), as opposed to coal (C)

During the 1990s, with deregulation of the power generation market in the United Kingdom, modern, high-efficiency, gas-fired combined-cycle plants replaced many old coal-fired power stations. This replacement meant that CO_2 production per MWh of electricity generated by these new power plants dropped to a third of the value for the coal-fired plants. This example shows that deregulation can have a strong positive ecological effect.

New concepts to reduce and even to avoid CO_2 emissions in power generation from coal will use combined-cycle plants as the major component of plants that first generate gas from coal by gasification followed by gas cleaning and the power generation process itself. These innovative Integrated Gasification Combined-Cycle plants (IGCC) are described in chapter 12.

Waste Heat Rejection

Another environmental concern is the waste heat that every thermal power station releases to the environment. Here too, the high efficiency of the combined-cycle plant is an advantage: from any given amount of primary energy, a greater amount of electricity or useful output is produced. This reduces the amount of waste heat released to the environment.

In addition to the quantity of waste heat, however, the form in which the heat is transferred to the environment is also important. The effect is less severe if the power plant heats air instead of releasing its waste heat to a river or the sea. Conventional steam power plants often dissipate the waste heat to water for efficiency reasons. The most economic solution for combined-cycle plants is frequently to dissipate waste heat to the air through a wet cooling tower. Direct air cooling is also possible, but results in a reduction in output and efficiency together with increased costs due to the air-cooled condenser (see chapter 7).

A combined-cycle plant needs only half the cooling water of a conventional steam plant of the same output and a third of what is required for a nuclear power station.

A gas turbine usually requires practically no external cooling except for the lube oil and generator. This fact has contributed greatly to its widespread acceptance in countries where water is scarce. Because the steam portion of the combined-cycle plant accounts only for a third of the output, an air-cooled condenser is an economic solution for dry areas, underscoring the wide range of possible applications.

Table 10–1 shows the amount of waste heat that must be dissipated for various types of plants with 1000 MW electrical output. All steam cycles are cooled by river water or seawater.

Table 10–1 Comparison of the heat to be dissipated for various types of 1,000 MW station

Heat sink	Gas turbine [MW]	Combined cycle [MW]	Steam plant [MW]	Nuclear plant [MW]
Air / stack	1,500–2,000	130–180	70–100	0
Water	0	550–700	1,100–1,400	1,800–2,000

Noise Emissions

An environmental issue that is considered during the design and construction of a combined-cycle power plant is noise. This issue can be resolved using acoustic insulation and silencing equipment available today.

A distinction is made between *near field noise* and *far field noise*. Near field noise refers to the noise levels at the machinery. Far field noise often based on the plant site boundary and indicates the noise emitted to the environs. Main sources for this noise are the gas turbine inlet, the gas turbine exhaust (or HRSG), the stack and cooling tower, or an air-cooled condenser. Steam bypass operation during startup and shutdown is an additional source of noise.

Conclusion

Due to their low emissions levels, low cooling requirements, and noise levels capable of meeting stringent requirements, combined-cycle plants are considered environmentally friendly and are well suited for decentralized power generation in urban areas.

11 Developmental Trends

New technology developments and operating modes of combined-cycle power plants are being driven primarily by economic needs and—with growing importance—by ecological constraints. At the end of the day these ecological constraints will also result in a commercial impact on the competitiveness of the respective technical concepts.

This means in general that new competitive combined-cycle power plants have to generate electrical energy at lower cost than competing power stations in the different market segments of base and intermediate load with their fluctuating electricity prices today and in the foreseeable future.

This can be achieved by plants that offer:

- High energy conversion efficiency
- Low investment and operating costs resulting in low life-cycle cost
- High reliability, availability, and maintainability
- Higher plant flexibility
- Improved maintenance concepts
- A broad range of applicable fuels and fuel flexibility
- Environmentally compliant operation and preparations for increasing environmental requirements

As a matter of course all these features have to be optimized from a commercial point of view to achieve the ability to compete.

Seven major developmental trends can be identified to meet the aforementioned objectives.

- Increased gas turbine firing temperatures to enhance the efficiency of the gas turbine, and simultaneously of the water/steam cycle thanks to higher main steam parameters
- New technical gas turbine concepts that will achieve a higher efficiency
- Increased gas turbine power output to benefit from the economies of scale
- Enhanced load cycle capabilities to enable intermediate load and frequency control operation
- Reduced operating costs by applying remote control and service packages
- Less emissions, especially NO_x, to reduce environmental impact
- The development of hydrogen-fired gas turbines as a prerequisite
  - for new plant concepts using other fuels than natural gas and oil
  - and/or to improve environmental compatibility (CO_2 capture), (see chapters 12 and 13).

Increased Gas Turbine Firing Temperatures

The positive effect of high gas turbine inlet temperatures on the efficiency of the combined cycle and the resultant reduction in the fuel cost component in the cost of electricity is pushing gas turbine inlet temperatures higher.

It seems reasonable to look for further improvements from even higher gas turbine inlet temperatures that have become possible through the development of new materials and improved cooling systems.

Research projects in this area focus primarily on improved air cooling of the hot-gas path of the gas turbine. Alternative cooling technologies employing steam cooling, for example, have been announced by gas turbine manufacturers and implemented in both prototype and commercial plants, but, due to the increased complexity, without the expected breakthrough.

One main constraint on raising the temperature is the allowable strength of the blades and vanes. One major effort concentrates on the strength of the base alloy and its crystal structure, going from conventionally cast low-nickel to high-nickel alloys with a single crystal. In the future even more important will be the development of the coating technologies, protecting the base alloy not only against corrosion but also against the temperature of the flue gas (thermal barrier coating or TBC). Some companies are considering ceramic structures (CMC) at least for the vanes. The respective development programs are ongoing but the results do not meet expectations. In particular the required mechanical strength and parts life remains a challenge for further rescarch. Figure 11–1 shows the increase of the gas turbine inlet temperatures over the course of time, and as a function of the specific development steps.

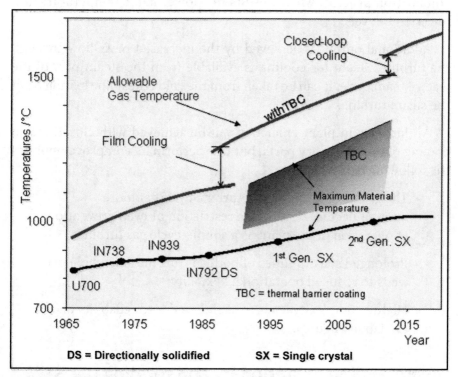

Figure 11–1 Chronology of the gas turbine inlet temperatures based on improved material and cooling technologies

Closed-steam cooling of stationary parts

The majority of gas turbines presently in operation have air-cooled blades and combustion chambers.

Cooling of the blading allows turbine inlet temperatures to be increased beyond the allowable material temperatures. If air cooling is employed, the cooling air first cools the blade internally and then exits onto the blade surface, providing a cooling film that separates the hot gas from the blade surface. However, mixing the two flows leads to a disturbance of the flow pattern and the cooling of the hot gas, which reduces efficiency. An alternative is to use steam as a cooling medium. Closed-steam cooling takes advantage of the fact that steam has a higher heat capacity than air, and doesn't deteriorate the expansion efficiency. Additionally, more air is available for the combustion process, which results in a lower flame temperature and lower NO_x levels.

Additional output is achieved by the increased mass flow through the turbine. Steam for cooling is available from the steam part of the water/steam cycle. It can be taken from the cold reheat line exiting from the steam turbine.

An increase in plant efficiency can be achieved with closed-steam cooling of the stationary parts, but this performance improvement has the following disadvantages:

- The gas turbine and steam turbine cycle get more interconnected, resulting in restriction or even prevention of operation of the plant as a simple-cycle gas turbine.
- Prolonged startup times of both gas turbine and whole plant leads to reduced operational flexibility.
- Higher complexity leads to slightly reduced reliability and availability figures.

Steam cooling of stationary and rotating parts

The improvement potential is even larger for steam cooling if the rotating parts are steam cooled as well.

A combined-cycle plant with a fully steam-cooled gas turbine is expected to offer higher efficiency and produces a larger specific power output per unit of inlet airflow. This could eventually result in lower specific costs, although maintenance costs are expected to be higher due to the higher blading costs. The following points are challenges for this type of gas turbine.

- Cooling of the thin front and rear ends of the blades
- Steam purity requirements for the cooling steam
- Leak tightness of the steam system (steam must be supplied to the gas turbine rotor and returned from there)

This type of plant is attractive from an economic point of view only if these challenges are resolved satisfactorily and the plant achieves the corresponding reliability. Also the disadvantages of steam cooling in general, mentioned in the previous paragraph should be taken in consideration. Whether the benefits of highly sophisticated and complex technology will be high enough to make it competitive is still an open question. The manufacturer is speaking about a *"...combined-cycle system capable of breaking the 60% efficiency barrier..."*

Larger compressors

Parallel to these developments, improvements are also being made to the compressor. The advantages offered by the higher gas temperatures can only be fully exploited if the pressure ratio of the machine is increased to an appropriate level. Increasing the airflow through the compressor also produces high unit ratings. With modern blading, compressors are able to handle volumetric flows that seemed utopian just a few years ago. With latest compressor models developed using modern design tools and incorporating long-term experiences with aero engines, it is possible to increase pressure ratios and reduce the number of compressor stages.

Higher main-steam parameters

Early combined-cycle plants had water/steam cycles with low steam parameters compared to conventional steam power plants. Main steam pressures in a range of 50 to 80 bar (710 to 1,150 psig) and main steam temperatures of 450 to 500°C (842 to 932°F) were standard.

Modern, large-capacity combined-cycle power plants for power generation operate at a main steam pressure of up to 170 bar (2,310 psig) and a main steam temperature up to 565°C (1,049°F), with the latest figure as high as 600°C.

For steam turbines beyond approximately 100 MW capacity, reheat machines are now standard compared to the previous non-reheat units. This was made possible by the sequential firing technology or higher firing temperature of the gas turbines, which, as a result, provided higher exhaust temperatures and, therefore, higher inlet temperatures to the heat recovery steam generator (HRSG). Higher gas turbine exhaust temperatures yield a higher optimal main steam pressure based on the net present value (NPV) of the plant. In the near future it can be expected that these trends will continue. Large plants for power generation will have higher main steam pressures, increased main steam temperatures, and reheat for the steam cycle.

The following points must be considered when the main steam parameters are increased.

- Higher main steam temperatures require more expensive alloys in the HRSG, steam piping, and steam turbine. The gain in output must, therefore, justify this additional investment.

- Higher main steam pressures cause wall thickness to increase, which, in general, reduces thermal flexibility and increases cost. Once-through HRSGs will be installed more frequently to avoid the negative impact on thermal flexibility of the higher main steam pressures.

- High main steam pressure in combination with a reheat steam turbine reduces the main steam volume, which may result in a reduction in the efficiency of the high-pressure steam turbine due to shorter blading.

The last mentioned effect would result in a main steam pressure that is too far away from the thermodynamic optimum. To avoid this relative disadvantage either two gas turbines and HRSGs are combined with one steam turbine to the typical 2 + 1 configuration, or the high-pressure steam turbine runs at higher speed, which reduces rotor diameter and allows an increase of blade length. A side benefit is the smaller casing dimensions, which improve thermal flexibility. Figure 11–2 shows a geared high-pressure turbine.

As the exhaust gas temperatures of the gas turbine rise further, the optimal cycle selection is affected.

The triple-pressure reheat cycle is often the optimal steam cycle for a gas turbine with an exhaust temperature of approximately 600°C (1,112°F). The optimal cycle will move towards a single-pressure reheat cycle for gas turbines with an exhaust temperature of 750 to 800°C (1,382 to 1,472°F).

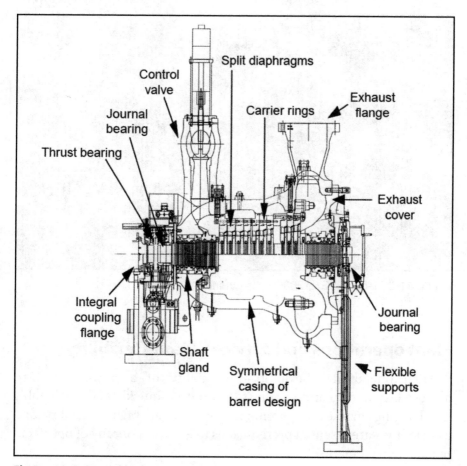

Figure 11–2 Geared high-pressure turbine

With these improvements combined-cycle plants with typical guaranteed net efficiency of around 58.5% are already in commercial operation, and the first modern plants achieving 60% are already in test and validation. New gas turbines will now be available with unit power capacities up to 340 MW for 50 Hz applications (figure 11–3) and 270 MW for 60 Hz applications leading to lower costs, so that combined-cycle plants will become more competitive for large power plants.

Figure 11–3 Siemens SGT5-8000H gas turbine

Plant operations and service—Plant flexibility

High operational flexibility—the ability of a power plant to achieve fast startup and to adjust load output quickly and predictably to changing market requirements, as well as high reliability and availability—are essential prerequisites to ensure economy of operation in a liberalized market.

Major factors limiting the load variation of an existing combined-cycle power plant are the allowed pressure and temperature transients of the steam turbine and the heat recovery steam generator, waiting times to establish required steam chemistry and warm-up times for the balance of plant and the main piping system. Those limitations also influence the fast startup capability of the gas turbine through required waiting times compared to a simple-cycle startup. Enabling combined-cycle power plants to improve operational flexibility requires a range of so-called fast cycling provisions. The most important ones are shown in figure 11–4, with the main emphasis on warm-up of the steam turbine with the first steam produced after start of the gas turbine and run-up during continuous pressure and temperature increase. As an example, these measures allow a startup time of less than 40 minutes.

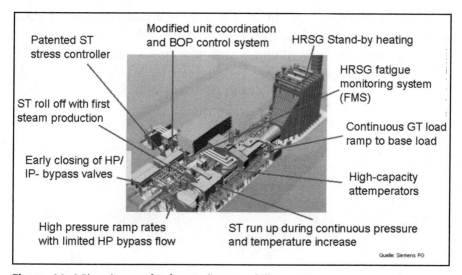

Figure 11–4 Plant impact for fast cycling capability

Maintenance and performance degradation optimization

The reliability and availability of a power plant are strongly dependent on an optimized operations and maintenance concept covering the whole lifetime of the plant. The original equipment manufacturers

(OEMs) offer a lot of service packages to support the power plant operator. Beyond the simple case-by-case or routine maintenance, refurbishment of the hot gas path parts, supply and installation of spare parts for components, these packages might be long-term and full-scope maintenance programs for components or the whole plant, and finally the complete operations and maintenance done by the OEM. To minimize the gas turbine performance degradation described in chapter 6, either the measures of the routine maintenance are applied or customized upgrade packages for the main components of the power plants, especially for the gas turbines, could be implemented. A classic example is the download of modern gas turbine technology into the older engines that have been operating for many years. That could be compressor upgrade to increase the air mass flow or upgrade of the hot gas path parts to make an increase of the firing temperature possible.

Incremental as well as disruptive technical improvements linked with aligned operational and service concepts are mandatory to ensure the competitive position of combined-cycle power plants in a market driven by changing economic and ecological challenges.

12 Integrated Gasification Combined Cycle

Environmental constraints as well as increasing prices of high-grade fossil fuels such as natural gas are major drivers that determine further development of fossil-fuel-fired power stations. One of the most attractive options to achieve extremely low environmental pollution is the Integrated Gasification Combined Cycle (IGCC). The IGCC concept, shown in figure 12–1, opens the well-proven combined-cycle concept to so-called *dirty fuels* such as coal, refinery residues, biomass, and wastes by adding gasification, air separation, and gas cleaning processes to the upstream gas turbine combustor.

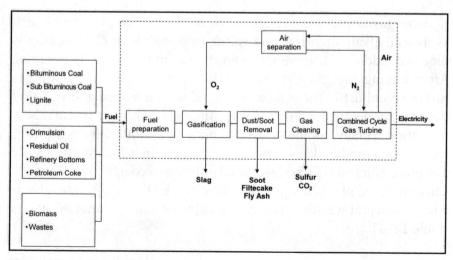

Figure 12–1 Principle IGCC concept

Gasification is basically a partial oxidation process that converts any carbonaceous feedstock such as coal, petroleum coke, heavy oil and oil tars, biomass, and waste streams into a gaseous product called synthesis gas (syngas). Syngas mainly consists of CO and H_2 (with some CO_2, H_2O, and contaminants), with the composition depending on fuel and type of gasifier (figure 12–2).

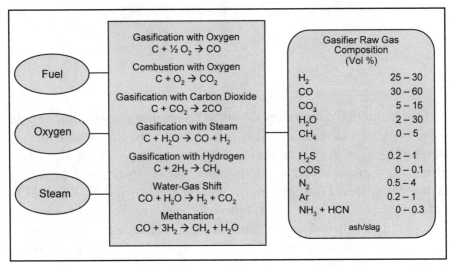

Figure 12–2 Gasification reactions

After gasification, the syngas is cleaned of its contaminants such as fly ash, alkali, chlorine, and sulfur compounds. If CO_2 capture is required, additional conversion and treatment steps can be foreseen. After leaving the gas cleaning section and before feeding the gas turbine combustor, the syngas is diluted with nitrogen and/or water vapor to moderate combustion condition and minimize thermal NO_x formation. Finally, combustion takes place in a combined-cycle power plant with a modified gas turbine. Such a system is inherently cleaner and more efficient than conventional pulverized coal (PC) fired plants. Current IGCC plants in operation have proven that syngas cleaning is a more efficient way than flue gas cleaning of conventional PC plants (table 12–1).

Table 12–1 Emission of existing coal based IGCC

IGCC	Start-up	PM [kg/MWhel]	NO$_x$ [kg/MWhel]	SO$_x$ [kg/MWhel]
Nuon Buggenum *The Netherlands*	01/94	0,005	0,318	0,200
Elcogas Puertollano *Spain*	12/97	0,020	0,399	0,068
Wabash River *USA*	10/95	0,045	0,494	0,490
Tampa / Polk Power *USA*	09/96	0,064	0,390	0,612
Advanced SPP	2010	0,030	0,530	0,400

Due to high investment cost, IGCC technology is primarily suitable for large centralized power plants with access to low-cost feedstock such as coal and petcoke. Moreover, petcoke with its high sulfur (>5% in weight) and vanadium content is suitable only for IGCC applications. Refinery residues such as heavy oils and tars are potential sources for hydrogen production used in refinery upgrading to produce diesel and petrol fuels. Therefore, refineries primarily focus on polygeneration concepts with hydrogen as the main product, whereas only surplus of syngas is used for electricity generation and export. Biomass and waste are best exploited using smaller plants close to their source or being co-gasified with coal in large IGCC plants to achieve better competitiveness.

The drawback of IGCC plants, however, is that they are significantly more expensive than conventional pulverised coal PC plants. Current IGCC plants in operation have suffered from relatively poor reliability and limited operational flexibility (e.g., long startup time). Nevertheless the demonstration plants built in the United States and Europe have proven that IGCC is a viable power generation technology (table 12–2). Improvements have been identified, and it is expected that further evolution of technology and commercialization will significantly reduce cost.

Table 12-2 Commercial solid fuel IGCC plants 2007

IGCC power plant	Buggenum	Wabash-River	Tampa	Puertollano	Vreshova
Location	Netherland	USA	USA	Spain	Czech Republic
Commissioning	1994	1995	1996	1998	1996 (2005)*
Net Power	253 MW	262 MW	250 MW	300 MW	351 (430) MW
Fuel/Coal	Hard coal/Biomass	Hard coal/Petcoke	Hard coal	Hard coal/Petcoke	Lignite
Gasifier	Prenflo	E-Gas (CoP)	GE	Shell	Sasol-Lurgi/SFG
Gasturbine	Siemens V94.2	GE 7FA	GE 7FA	Siemens V94.3	GE 9FE
Net Efficiency (LHV)	43 %	39 %	41 %	42 %	44 %

*Plant extension 2005

Gasification Technologies

Gasification became a commercial process in 1812 and was initially used to produce town gas for lighting and heating purposes. Since the early 1900s syngas has also been used as a chemical feedstock. Development of today's large-scale gasification processes took place in the 1920s to the 1940s, mainly in Germany.

In a gasification reactor, first drying of the feedstock particles takes place. As the temperature rises, devolatilization occurs and tars, oils, phenols, and hydrocarbons are formed. With further increased temperature these devolatilization products are decomposed to lighter products such as CO, H_2, and CH_4. The fixed carbon, which remains after devolatilization, is gasified via reactions with O_2, H_2O, and CO_2. Beyond hydrocarbons other fuel constituent elements such as sulfur, nitrogen, and ash are of major consideration because they also react and their reaction products (e.g., H_2S, COS, HCN, NH_3, alkaline, chlorine, slag, and ash) have to be removed either out of the gasifier or downstream in a gas cleaning process.

Since the 1920s, three basic gasification technologies have been developed that are applicable for IGCC in principle: fixed-bed processes, fluidized-bed (FB) processes, and entrained-flow processes. All types of gasifier can be designed to operate with air or pure oxygen (>85%)

as gasification agent. Oxygen-blown gasification offers a number of advantages over air-blown systems because they promise a higher cold gas efficiency and carbon conversion rate and smaller plant component dimensions due to absence of nitrogen surplus. Gasifier technologies differ in temperature operating range, carbon conversion rate, particle size, and hydrocarbon content in raw gas starting from fixed-bed to entrained-flow gasification. A summary of technologies is given in figure 12–3.

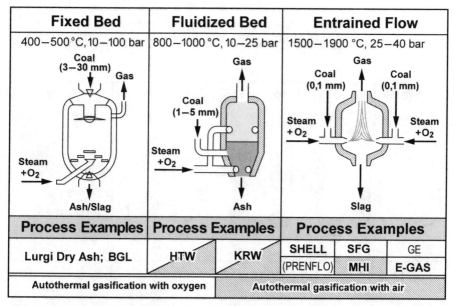

Figure 12–3 Basic technologies of gasification and gasifier technology vendors

Fixed-bed gasification is characterized by significant steam requirements for temperature control, long carbon conversion time, and a high content of higher hydrocarbons in the product such as tars and oils. The process is, therefore, not preferable for IGCC applications. Modern large-scale IGCC applications are based on FB or entrained-flow gasification concepts. FB gasification is characterized by gasification temperature below the ash melting point and subsequently lower oxygen consumption, which leads to a higher IGCC efficiency. Moreover as a consequence of lower gasification temperature, carbon conversion rate reduces resulting in extensive char recycle and significant carbon loss. FB gasifiers are, therefore, suited only for high volatile coals such as lignite or Powder River Basin (PRB) sub bituminous coal.

Entrained-flow gasifiers operate with gasification temperature above ash-slagging conditions, achieve carbon conversion rates higher than 98%, and offer very short residence time leading to highest unit capacity. Although oxygen consumption is higher and IGCC efficiency lower compared to FB gasification, entrained-flow gasification systems are compatible with almost all kind of liquid and solid hydrocarbons.

There are numbers of solid-based entrained-flow gasifier types available, which differ in various features. They are summarized in figure 12–4 and explained in more detail following the following sections.

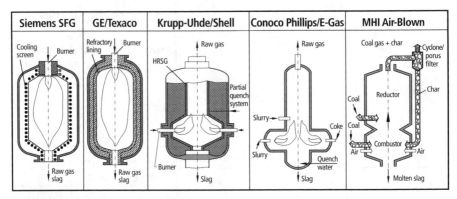

Figure 12–4 Entrained-flow gasifiers

Shell coal gasification process (SCGP)

- Single-stage gasification with dry feed
- O_2 blown
- Reactants flow upward and molten slag downwards; reactor employs a vertical tube membrane wall for cooling that generates IP steam
- Syngas cooling done by recycle of cooled syngas and further cooling by HP/IP steam generation in water tube cooler design

Advantage: Highest IGCC efficiency due to dry feed and heat recovery with HP/IP steam generation

Disadvantage: High investment caused by high alloy heat exchanger materials to prevent corrosion

GE coal gasification process (former Texaco)

- Single-stage gasification with slurry feed
- O_2 blown
- Reactants and molten slag flow downwards; reactor is equipped with refractory (ceramic brick liner)
- Syngas outlet temperature reduction is done by either full water quench or radiant and convective fire tube cooler

Advantage: Extensive commercial operating experience

Disadvantages: Limited lifetime of refractory and burner
Long startup time due to need of refractory preheating
Limited ability to handle low rank coals
Relatively high oxygen consumption due to slurry feed

Siemens fuel gasifier (former GSP/Future Energy)

- Single-stage gasification with dry feed
- Reactants and molten slag flow downwards; reactor employs a spiral-wound coil-cooling screen
- Syngas cooling done by full and potentially also partial water quench; in case of partial water quench convective heat recovery for coal under development

Advantages: Low specific investment and oxygen consumption
Proven for low-rank coals and most suitable for IGCC with CO_2 capture due to water quench
Fast startup

Disadvantage: Limited operational experience up to 200 MW thermal throughput
Lower efficiency due to quench cooling

ConocoPhillips E-Gas gasification process

- Two-stage gasification with slurry feed—first stage accomplishes exothermic oxidation and second stage provides syngas temperature reduction to 300–350°C by endothermic chemical and water quench (slurry); further cooling by fire tube syngas cooler
- Char/solid recycle to first stage
- Reactants flow upwards and molten slag downwards; reactor employs a refractory wall

Advantage: Relative high IGCC efficiency due to chemical quench and heat recovery

Disadvantages: Limited lifetime of refractory and burner
Long startup time due to need of refractory preheating
Limited ability to handle low-rank coals
Limited suitability for CO_2 capture due to high methane content in syngas

MHI dry-feed gasifier

- Air-blown gasifier
- Two-stage gasification with dry feed—first stage is exothermic oxidation and second stage is chemical quench by endothermic chemical reaction
- Reactants flow upward and molten slag downwards; reactor employs a cooled membrane wall
- Syngas cooler

Advantage: High IGCC efficiency

Disadvantage: Technology in demonstration phase, no commercial reference yet

Syngas Cleaning Technologies

The syngas cleaning process is required to remove compounds, which harm the environment and/or lead to corrosion and thus a limited lifetime of components. Alkali compounds are especially responsible for high-temperature gas turbine corrosion and, to a very high extent, have to be removed. Table 12–3 provides gas turbine limitation for chemical impurities and dust.

Table 12–3 Allowed chemical impurities of gas turbine fuels (Siemens source)

Limits for chemial impurities (Fuel Weighting factor f=1)				downstream filter
Pollutant		Test/Check	Unit	Syngas
Dust (with natrural gas), sediments (with EL distillate Fuel)	Total $d < 2$ µm $2 < d < 10$ µm $d > 10$ µm $d > 25$ µm	DIN EN 12622 VDI 2066 (1994) ASTM D 2709 & D 6304	ppm (wt)	< 20 < 18.5 < 1.5 < 0.002
Vanadium (V) (in case of fuel oil)		DIN 51790 ASTM D 3605	ppm (wt)	< 0.5
Lead (Pb) (in case of fuel oil)		DIN 51790 ASTM D 3605	ppm (wt)	< 0.1
Total of Sodium (Na) + potassium (K)		EPC method DIN 51790 ASTM D 3605	ppm (wt)	< 0.5 < 0.3
Calcium (Ca)		ASTM D 3605	ppm (wt)	< 10
Hydrogen sulfide (H_2S)		ASTM D 6228	ppm (vol)	< 100

Limits for fuel impurities (dust and ash, V, Pb, Na, K, Ca) are based on a lower heating value (LHV) f 42,000kJ/kg.

Formula X = LHV/42 [X= pollutant content; LHV in MJ/kg] shall be used to correct for deviations in lower heating value

To treat and purify the syngas leaving the gasifier a number of cleaning steps are needed, starting with removal of soot, char, and ash particles. This is followed by halogen, ammonia, and alkaline removal, and finally with sulfur, mercury, metal carbonyl, and, if required, CO_2 capture.

Dedusting

The particulates removal as first step can either be done by water scrubbing or using dry filtration methods such as cyclones or ceramic candle filters. In case of high particle flow and significant unconverted carbon (char) containing syngas as it occurs in two-stage entrained-flow gasifiers (E-Gas, MHI) or FB (HTW), a dry removal and char recirculation into the gasifier reaction zone is applied. Bulk removal of char and fly ash is done by cyclones or/and ceramic candle filter systems, which are sensible against plugging but require regular cleaning by flushing. The water Venturi scrubber operates at syngas dew point and also washes some soluble vapors such as ammonia, HCN (hydrogen cyanide) halogen, and alkali contaminants. Therefore, current gasification systems employ water scrubbers for final mechanical cleaning and removal of soluble compounds. Gasifiers with full-water quench such as GE and Siemens apply water scrubbing only (figure 12–5), whereas gasifiers with heat recovery such as Shell (figure 12–6), E-Gas, FB, and so on normally use dry fly ash/char removal steps and water scrubbing for final dust and ammonia removal. Full quench systems, therefore, are simpler with regard to particle removal.

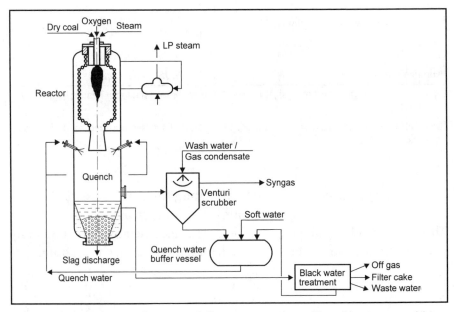

Figure 12–5 Schematic of Siemens full water quench gasifier with water scrubbing

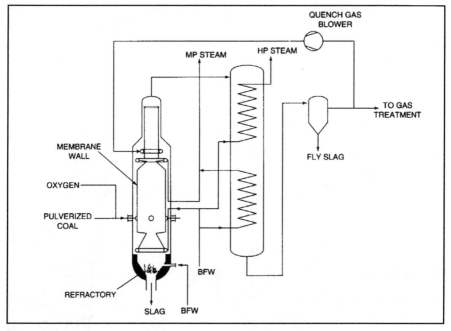

Figure 12–6 Schematic of shell coal gasification process with heat recovery and dry fly ash removal

Acid Gas Removal (AGR)

The next step in the syngas cleaning chain is acid gas removal to take out sulfur compounds such as H_2S (hydrogen sulfide) and COS (carbonyl sulfide), and if required also CO_2. In case of CO_2 removal an additional catalytic shift reactor is needed to convert CO into CO_2.

AGR systems currently employed in IGCC application are regenerable solvent-type processes (figure 12–7). The following three types of processes are known.

- Chemical solvent processes (e.g., Methyldiethanolamine (**MDEA**). or ADIP)
- Physical solvent processes (e.g., Rectisol, Selexol)
- Mixed chemical/physical solvent processes (e.g., FLEXSORB)

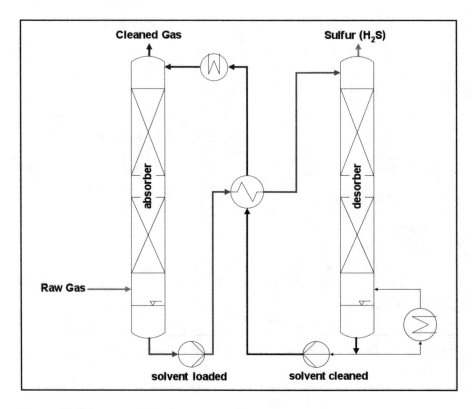

Figure 12–7 Regenerable solvent-type AGR process

AGR processes for IGCC application without CO_2 capture target to leave as much CO_2 as possible in the syngas to maximize the gas turbine syngas flow and power output. Thus, high selectivity for H_2S and COS is needed. Numerous studies concluded that IGCC application favors processes using MDEA solvent as a low capital cost option. Because MDEA is a selective H_2S removal process only, a COS hydrolysis (catalytical conversion to H_2S) needs to be applied before.

$$COS + H_2O = H_2S + CO_2 \text{ (exothermic)} \qquad (12\text{–}1)$$

An alternative promising AGR process applied in several gasification projects is the SELEXOL process. SELEXOL as a physical solvent process is more effective with increasing syngas pressure because its solubility directly depends on partial pressure and follows Henry's law.

The sour gas (H_2S enriched) leaving the desorption column of the AGR process is further processed in sulfur recovery and tail gas treating units. Sulfur recovery involves H_2S conversion into elemental sulfur or alternatively to sulfuric acid (H_2SO_4). However, the most common technology for sulfur recovery is the Claus process that produces liquid sulfur by means of multiple catalytic conversion. Today air- and oxygen-blown Claus processes are used. If the H_2S feed content is low, modified Claus processes such as CLINSULF can also be applied. The overall chemical Claus reaction is:

$$3H_2S + 3/2O_2 = 2H_2O + 3S \text{ (exothermic)} \qquad (12\text{--}2)$$

A simplified Claus process is shown in figure 12–8.

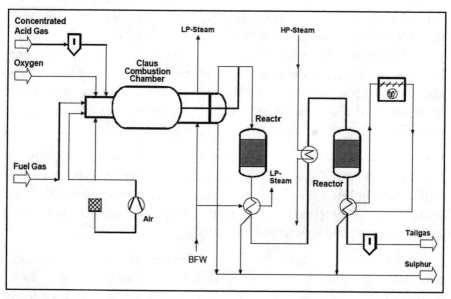

Figure 12–8 Claus process

Because Claus plants typically have sulfur recovery rates of less than 99%, the tail gas leaving the process still contains some sulfur and needs further processing. Depending on the sulfur emission level allowed, it can be fed to an incinerator, recycled upstream of the COS hydrolysis unit after hydrogenation of the Claus tail gas, or converted via additional hydrogenation/hydrolysis units.

Alternative and Advanced Gas Cleaning Technologies

Alternative processes to AGR and subsequent tail gas treating are direct oxidizing desulfurization methods such as SulFerox or LO-CAT. Both technologies use iron as oxygen carrier to directly convert H_2S into elemental sulfur, which is then filtered to produce sulfur filter cake.

Further technology development for syngas cleaning focuses on warm gas desulfurization processes, which target higher IGCC efficiency and cost reduction. These processes operate at temperatures of 400 to 500°C, which means that the syngas-cooling step can be limited. It allows feeding the gas turbine with syngas temperatures well above 400°C. The hot and warm gas reactor of such processes makes use of H_2S adsorption with metal oxide sorbents such as Zn or Ni. For continuous operation the reactants have to be regenerated by contact with air in a separate vessel. Although these developments have been tested successfully for sulfur removal, other contaminants (e.g., alkalis) are difficult to capture, and harm the gas turbine lifetime.

CO_2 capture becomes more and more important and leads to zero-emission IGCC concepts (ZEIGCC). CO_2 capture is done in two steps, shifting of CO to CO_2 and subsequently capturing the CO_2 by using AGR solvent processes. Future ZEIGCC concepts could also use more advanced membrane systems.

Depending on capture rate and selected gasification technology, two different shift reactor systems are commonly applied: high temperature sour gas shift (HT sour shift) or low temperature sweet gas shift (LT sweet shift). HT sour shift reactors achieve a CO conversion rate down to about 1 to 5 vol% residual CO, and LT sweet shift reactors down to about 0.05 to 0.5 vol% residual CO. Although LT sweet shift reactors are more efficient, their catalyst price is higher and operation much more complex. Thus, LT sweet shift reactors are applied only if a high carbon capture rate is required, or if the syngas has high methane content that cannot be converted to CO_2.

The water gas shift reaction is described as:

$$CO + H_2O = CO_2 + H_2 \text{ (exothermic)} \qquad (12\text{-}3)$$

An H_2O to CO ratio of about 1.1 to 1.3 is required to convert CO to CO_2. The water gas shift reaction is an exothermic reaction producing sensible heat, which can be used for HP or IP steam generation. Hence chemical energy is disposed causing lower gas turbine throughput and significant efficiency losses compared to standard IGCC processes. The conversion steam can either be introduced by full water quench (GE, Siemens gasifier) or additional IP steam injection or syngas saturation. For IGCC with CO_2 capture entrained-flow gasifiers with water quench are the preferred solution because of simpler process design and lower investment cost.

After-shift reaction, the syngas consists mainly of H_2 and CO_2, and the CO_2 can then be removed by a standard AGR process. It has been shown that CO_2 capture can most preferably be done by simultaneous sulfur and CO_2 removal using a physical absorption process (figure 12–9) such as Rectisol or Selexol. These processes consist of one column for CO_2 absorption and a second column for sulfur removal. The desorption process separates the H_2S/COS from CO_2 and feeds the sulfur gas to the Claus plant. The CO_2 is sent to further compression or liquefaction units.

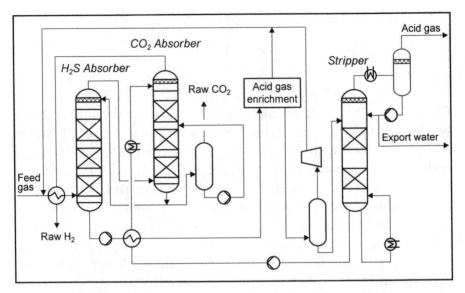

Figure 12–9 Physical absorption process for simultaneous sulfur and CO_2 removal

For IGCC application with CO_2 capture the Rectisol process is mainly considered today. The Selexol process shows a number of similarities to Rectisol, but leads to lower capture rate and an additional COS hydrolysis step caused by relatively poor selectivity between COS and CO_2. Another process available for CO_2 capture is an activated MDEA.

Air Separation Technology

Nowadays, almost all IGCC concepts are based on oxygen-blown gasifiers because they offer a number of advantages. The oxygen is supplied by an air separation unit (ASU). The oxygen is mainly used for gasification, and the pure nitrogen is used for feeding, pressurization, and flushing after shutdown. For IGCC cryogenic air separation units are used. Typical O_2 purities are approximately 95 vol% and N_2 purities are >99 vol%. Remaining nitrogen is used for syngas dilution and control of gas turbine NO_x emission. The cryogenic ASU (figure 12–10) consists of a cold box where oxygen and nitrogen are separated, a molecular sieve to remove CO_2 and water vapor and for downstream compression an oxygen pump and nitrogen compressor station. The cryogenic air separation process operates at temperatures levels lower than −150°C and can be designed for different pressure levels depending on gas turbine/ASU integration.

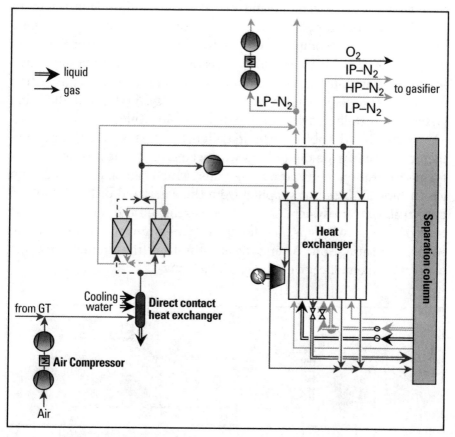

Figure 12–10 ASU process

Large-scale cryogenic air separation units are commercially proven and optimized, but represent the largest consumer of auxiliary power consumption within an IGCC application. Thus alternative air separation processes such as high temperature oxygen or ion transport membrane systems (OTM, ITM) are under development to minimize power consumption and to also reduce IGCC investment. Expected commercialization is onwards 2010.

Combined-Cycle Technology and Integration of Gas Turbine and Air Separation Unit

The combined cycle part of an IGCC is normally based on a three-pressure reheat cycle with additional water/steam side interfaces to a gasifier, gas cleaning, and air separation unit. Differences to standard combined-cycle configurations can be found in gas turbine modifications to accomplish compressor air bleed, higher turbine mass flow, and additional systems for syngas dilution and preheating. These adaptations are introduced to minimize NO_x emission and combustion instabilities but also to maximize IGCC efficiency. A syngas conditioning system is installed prior to the gas turbine inlet (figure 12–11). This system has the objective to first dilute the syngas with nitrogen and then to humidify the diluted gas with water vapor. Heat is needed for syngas saturation to evaporate the water, which can be recovered from gas turbine air bleed or supplied from the gas island or the combined cycle water/steam system. If air bleed is selected the sensible air heat is recovered in two steps, first for steam generation and second as heat supply for saturation. After syngas saturation, the diluted syngas is preheated to enhance overall IGCC efficiency.

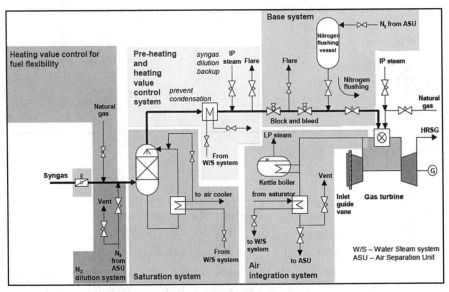

Figure 12–11 Gas turbine syngas conditioning system (Siemens concept)

Syngas dilution is primarily required for NO_x control. For this purpose, water vapor reduces flame temperature peaks and NO_x emission more efficiently than nitrogen because of its higher specific heat capacity (figure 12–12). The level of syngas saturation has a direct impact on IGCC efficiency and power output, which can be maximized by low temperature heat utilization through adding water of temperature between 150°C to 200°C up to a water content of approximately 10% (mass in syngas). In case a high syngas water content is required to reduce NO_x emission, the water temperature supplied to the saturator needs to be increased, which in turns lowers steam turbine power output and plant efficiency.

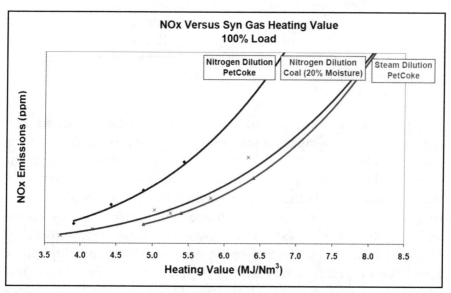

Figure 12–12 Effect of dilution and heating value on NO_x emission

After syngas dilution, saturation, and preheating, the syngas is fed to the gas turbine combustor. Compared to natural gas combustion, syngas combustion differs in various parameters, which require gas turbine modification with special focus on combustor design (figure 12–13). Advanced natural-gas-fired gas turbines are designed with pre-mix combustion systems to minimize thermal NO_x emission, but syngas contains significant amounts of hydrogen that cause higher flammability and risk of flashback and burner overheating.

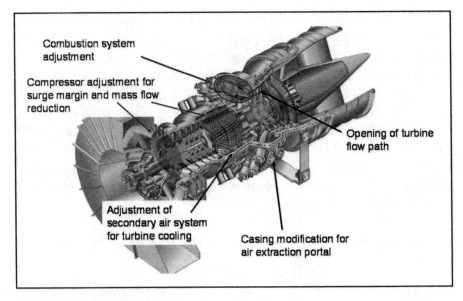

Figure 12–13 GT modification for syngas application

For these reasons, diffusion-based combustion systems are currently used for syngas utilization. Syngas pre-mix combustors are under development and might be available for the next generation of gas turbine applications. Although diffusion combustors offer higher flame stability than pre-mix combustors, the fuel/air mixture is inhomogeneous and leads to higher combustion temperatures and NO_x formation, respectively. Syngas dilution helps to bring NO_x emissions down to <25 ppm. By dilution, syngas heating values are in the range of 4 to 10 MJ/kg, depending on diluent type and turbine inlet temperature. Thus, large fuel pipes and flow control valves as well as an increase of turbine mass flow and corresponding imbalance of compressor and turbine section compared to the standard natural gas design have to be considered (figure 12–14). The higher turbine mass flow leads to a higher pressure drop over the turbine section and a higher compressor exit pressure.

Chapter 12 Integrated Gasification Combined Cycle • 307

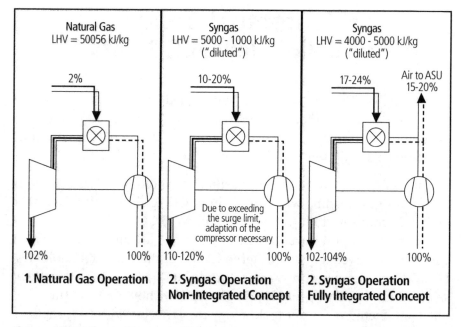

Figure 12–14 Gas turbine mass flow imbalance

With increasing compressor exit pressure, the surge limit of gas turbine compressors is approached or can even be exceeded. Under these conditions the risk of abrupt reversal airflow of compressor blades increases, which would harm the gas turbine compressor section. Beside surge limit, the increasing compressor exit pressure also affects the pressure profile of the gas turbine compressor and has an impact on the gas turbine cooling system. Therefore, compressor or turbine modification is required; for example, adding stages and staggering critical stages to enable high pressure ratio at lower mass flow, or opening the first turbine vane stage to reduce compressor exit pressure at given turbine mass flow. Alternative solutions to overcome these modifications are part load operation with throttled inlet guide vanes or air bleed to ASU at compressor exit. This has been done for both European IGCC plants at Buggenum and Puertollano where the gas turbine compressor supplies all the air needed in the air separation unit.

Conceptual Design of Solid Fuel IGCC Plants

Although IGCC application can be designed for various solid or liquid fuels and can be used for electricity generation only or co-production of chemicals, the main market is expected to push IGCC applications based on coal and used only for power generation.

Differences not only among gasifier technology but also various options of gas cleaning with and without CO_2 capture offer a number of IGCC concepts today. Four exemplary IGCC concepts with and without CO_2 capture are discussed in the following section, which covers coal diversity and differences in gasification technologies. HTW (high temperature winkler) and Shell gasification have been selected for demonstrating IGCC concepts without CO_2 capture, and Shell and Siemens gasifier technology for concepts with CO_2 capture.

IGCC Concept Based on Hard Coal and Shell Gasification and Siemens Combined-Cycle Technology

The Shell-based IGCC plant concept has been derived from the existing demonstration plant at Buggenum, but improved by using advanced gas turbine F-class technology as well as MDEA gas cleaning and partial air-side integration between ASU and gas turbine to enhance operational flexibility (figure 12–15). Heat and mass balance calculation are based on ISO ambient conditions, and a typical hard coal such as Douglas Premium is assumed.

The hard coal must be milled and dried before entering a Shell gasification process where it is converted to syngas. The hot syngas leaving the Shell gasification reactor is quenched with recycled raw gas down to about 800°C to 900°C and subsequently further cooled to about 250°C while generating high-pressure and intermediate-pressure steam. Particle removal takes place by use of ceramic filter units, and the raw gas is further processed by Venturi scrubbing to remove alkali and halogen contaminants. About 80% of the initial chemical heat

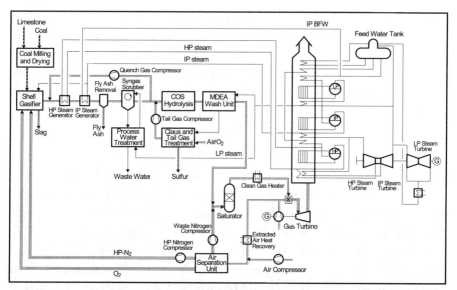

Figure 12–15 IGCC concept based on hard coal and Shell gasification and Siemens combined cycle technology

content of the feedstock is still present in the syngas after gasification and first cleaning steps are completed. Part of the chemical energy is converted to HP/IP steam and used for power generation in a steam turbine. Although this energy is still used for power generation, the energy conversion is less efficient because it is used only at the lower temperature level of the Rankine cycle and is not using the high temperature potential of the full combined cycle. The heat losses of the entire gasification island are present in fly ash, slag, and cooling water, and have been estimated to about 2.6%.

After water scrubbing, a combination of COS hydrolysis and MDEA has been assumed to remove sulfur compounds. The sour gas leaving the MDEA is treated in a conventional Claus plant and tail gas hydrogenation unit. Energy losses of entire sulfur removal step are estimated to about 3.1%.

The oxygen needed for Shell gasification process is supplied by a conventional ASU process, which is integrated with the gas turbine on air and waste nitrogen side. All the waste nitrogen separated from compressed air returns for syngas dilution, and about 50% of the air input is supplied by the gas turbine compressor leading to a partially integrated concept. Energy losses by ASU have been estimated to about 5.3% and belong to cooling water consumption.

The diluted syngas enters the gas turbine combustor where the remaining chemical heat is converted to electricity and sensible heat of leaving flue gas. Based on original chemical heat input of coal the gas turbine efficiency can be estimated to about 27.3%, assuming a typical F-class engine with turbine inlet temperature of 1230°C (ISO). The flue gas leaving the gas turbine has a temperature of more than 600°C and is used for steam generation within a triple-pressure reheat HRSG configuration. The HP and IP saturated steam imported from the gasification island is fed to the HP and IP drum, and jointly superheated with internal generated steam fraction. To accommodate the saturated steam import from gasifier island, the HP/IP superheater section is oversized for secondary fuel operation and, therefore, higher water consumption for steam attemperation is needed if the gas turbine is operated with natural gas or fuel oil. In coal gas operation the steam turbine power output is about 45% higher due to steam generation in gasifier heat recovery and results to approximately 19.1% relative to entire chemical heat input (figure 12–16). To achieve the highest IGCC efficiency level while limiting NO_x emission, the gas turbine is integrated with the air separation unit on the air and nitrogen side. Nitrogen dilution is used to reduce NO_x emission, but increases turbine mass flow. To compensate the resulting turbine imbalance, part of compressed air is extracted and supplied to the air separation plant.

IGCC Concept Based on Lignite and HTW Gasification and Siemens Combined Cycle

The most mature pressurized FB technology is the high temperature winkler (HTW) process, which was tested on demonstration scale at Berrenrath, Germany. Because of its high efficiency potential overall performance estimations based on ISO condition, a blended German

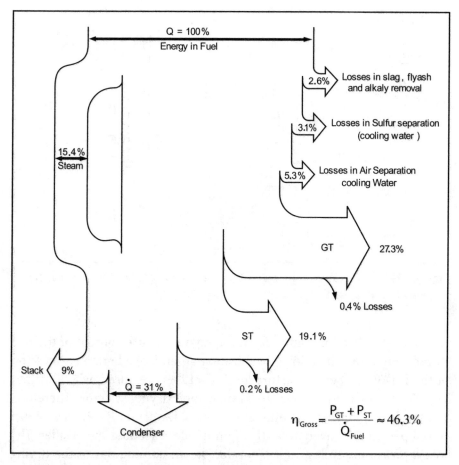

Figure 12–16 Sankey diagram of Shell-based IGCC concept

lignite coal and the Siemens combined-cycle technology have been performed. Although more efficient than entrained flow IGCC application, the FB technology is characterized by a limited carbon conversion rate and a carbon-containing ash leaving the bottom of the gasifier (figure 12–17). The bottom product cannot be disposed of and is used for HP steam generation in an additional fluidized bed combustor. The superheated steam is exported to the HP steam section of the combined cycle HRSG.

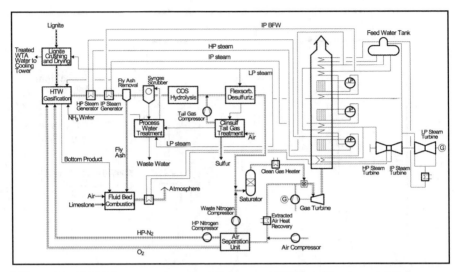

Figure 12–17 IGCC concept based on lignite and HTW gasification and Siemens combined cycle

Lignite applications suffer from the high moisture content of the fuel of typically more than 54%. Establishing an effective lignite gasification at the HTW gasifier a moisture level of <12% is required, which cannot be achieved by conventional coal milling and drying systems. Therefore, fluidized bed drying systems such as WTA from RWE have been developed to effectively treat the lignite before feeding the gasifier. The WTA process recycles the evaporated steam coming out of the drying process for heat input into the system after compression. For balancing energy losses, LP steam from combined cycle is used and integrated. The crushed and dried lignite leaving the WTA unit enters the fluidized bed of the gasifier and reacts with oxygen and IP steam at about 1000°C but below ash melting point. The produced hot syngas leaves the reaction zone and contains still significant amounts of carbon and fly ash, which are separated in a cyclone and re-injected into the fluidized bed. After coarse ash removal, the syngas is cooled down to about 250°C in a heat recovery generator where saturated HP/IP steam is produced. Downstream of the syngas cooler, an additional fine ash removal filter is situated to separate the remaining fine ash from the syngas stream. Total losses of coal conversion and mechanical cleaning process are about 1.4%, and 10.6% of chemical energy is converted to steam. Further treating steps are needed after syngas cooling and de-dusting, starting with water scrubbing to remove trace solids, ammonia, and acids. Afterwards the syngas enters the COS hydrolysis unit and COS is converted into H_2S, which than is removed by a FLEXSORB SE+ absorption process. FLEXSORB SE+ represents a very selective chemical solvent well suited to remove low H_2S concentration. After desorption the H_2S is fed to a CLINSULF reactor (comparable to Claus process; see figure 12–8) and converted to elemental sulfur while the off gas leaving the CLINSULF unit is converted back to H_2S by hydrogenation and returns upstream to the FLEXSORB SE+ unit. Beyond sulfur removal the FLEXSORB unit also absorbs naphthalene, which is formed in the HTW gasifier due to its lower gasification temperature. The total energy losses of sulfur removal amounts to about 1% and of air separation to about 3.3%.

The combined cycle in case of HTW gasification represents a highly integrated system of importing and exporting steam and water on different pressure levels and temperature condition. Although the entire IGCC net efficiency level is higher than for Shell gasification (see table 12–4), the efficiency benefit is compensated by higher investment and system integration in case of lignite application. Gas turbine integration on the air and nitrogen side has been considered similar to the Shell case.

Table 12–4 Investment and efficiency of IGCC concepts

	Shell based IGCC (hard coal) (50 Hz)	HTW based IGCC (Lignite) (50 Hz)	Shell based IGCC with CO_2 capture (50 Hz)	SFG based IGCC with CO_2 capture (50 Hz)
Net power output	874 MW	826 MW	737 MW	681 MW
Gross power output	986 MW	931 MW	956 MW	864 MW
Gasification island	36%	36%	32%	23%
Gas cleaning	6%	9%	13%	15%
ASU	7%	5%	7%	8%
Combined Cycle	28%	27%	24%	28%
BoP and Owners cost	23%	23%	24%	26%
Spec. total costs	1500 €/KW	1600 €/KW	2050 €/KW	1850 €/KW
Net Efficiency (LHV)	46,3%	51,7%	35,9%	34,5%
Net Efficiency (HHV)	44,5%	43%	34,4%	33,1%
Coal	Douglas Premium 2	Lignite	Douglas Premium 2	Douglas Premium 2

IGCC Concept with CO_2 Capture Based on Hard Coal and SFG Gasification and Siemens Combined Cycle

The SFG technology is currently based on a dry feeding system and a full water quench for syngas outlet cooling to about 200°C and leads to a syngas water content of more than 50 vol% when the syngas leaves the gasifier quench section. SFG gasification with partial water quench system and heat recovery can also be applied for future application presently under development, which in general leads to higher IGCC efficiency levels. But if CO_2 capture is required more than 50% syngas water level is needed to perform the water gas shift reaction from CO into CO_2 and H_2. In this case a full water quench solution such as SFG gasification offers the most economical and simplest IGCC configuration. No or limited steam injection is needed to accommodate the water gas shift reaction and the Siemens IGCC concept with CO_2 capture comprises only few interfaces between the combined cycle and gas island, which are shown in figure 12–18.

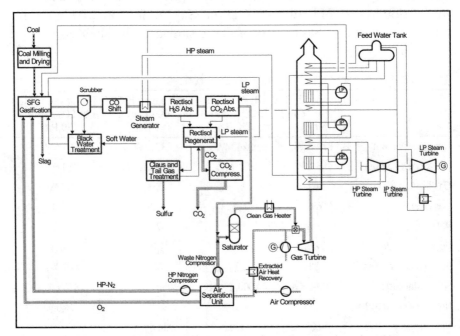

Figure 12–18 IGCC concept with CO_2 capture based on hard coal and Siemens fuel gasifier (SFG) and Siemens combined-cycle technology

ISO ambient condition, a typical hard coal such as Douglas Premium, and 90% CO_2 capture have been assumed for performance estimation of an IGCC concept based on SFG technology. The CO_2 is assumed to be compressed to about 110 bar.

Before gasification the hard coal is milled, dried, and fed to the gasifier where the coal reacts with oxygen and steam under lambda values of 0.4 to 0.5. After gasification and water quench, the syngas is cleaned by water scrubbing to remove particles and acids, and is then fed to a CO raw gas shift reactor to convert CO into CO_2 and H_2 by exothermic water gas reaction. After leaving the shift reactor, the syngas contains about 42 vol% H_2, 30 vol% CO_2 and still 24 vol% H_2O which depends on water/gas ratio and uses catalyst material (iron or cobalt molybdenum). The syngas temperature rises during shift reaction to more than 500°C and HP steam can be generated and exported to the combined cycle. After shift conversion, the CO_2, H_2S, and COS as well as other syngas traces are removed by a two stage Rectisol process. The sulfur compounds are absorbed in a first column and converted to elemental sulfur in an OxyClaus process followed by a tail gas treatment—hydrogenation—plant as sulfur recovery process. The CO_2 is removed in a second column, separated from solvent (methanol) by flashing on lower pressure levels, and compressed for transport and sequestration. The syngas leaving the Rectisol process contains of about 85 vol% hydrogen and needs to be diluted with nitrogen and water vapor (saturator) to control NO_x emission and reduce flame speed during gas turbine combustion. To enhance IGCC efficiency an additional syngas preheating step is assumed before the hydrogen-rich syngas enters the gas turbine combustor. Main energy losses are caused by CO conversion, CO_2 removal, and compression. The entire IGCC net efficiency level is estimated to be about 34.5% based on F-class gas turbine. Integration of gas turbine and air separation unit are considered to be comparable to IGCC cases without capture. The water steam cycle is based on a three-pressure reheat with only few interfaces to gasification and gas cleaning. Thus the combined-cycle operation does not change between syngas/hydrogen and secondary fuel operation.

IGCC Concept with CO_2 Capture Based on Hard Coal and Shell Gasification and Siemens Combined-Cycle Technology

The Shell gasification technology for IGCC with CO_2 capture corresponds to the standard Shell application, and performance estimations are based on ISO ambient condition, a typical hard coal, and 90% CO_2 capture with CO_2 compression to about 110bar (figure 12–19). The syngas leaves the Shell gasification island after water scrubbing and contains than about 9 vol% water vapor, 57 vol% CO, and 23 vol% hydrogen, which needs to be converted to hydrogen and CO_2 for downstream CO_2 capture. To accommodate the water gas shift reaction the syngas needs to be saturated first to about 50 vol% water vapor. This can be done by adding a cooler/saturator cycle that transfers the heat for water evaporation from shift reactor outlet to inlet side and by additional IP steam injection to cover the remaining steam requirement. Thus the sensible heat generated during exothermic water gas reaction is used efficiently, but significant investment and higher plant integration need to be considered, too.

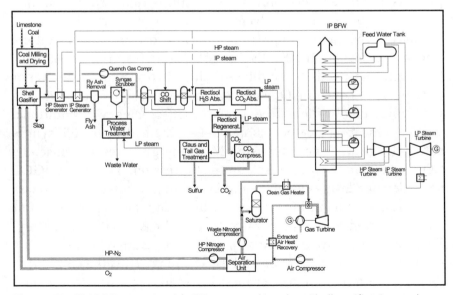

Figure 12–19 IGCC concept with CO_2 capture based on Shell gasification and Siemens combined-cycle technology

The shifted gas is then fed to a Rectisol process where CO_2 and sulfur compounds are removed in two separate columns. The process design of gas cleaning and conditioning corresponds to the SFG concept consisting of dilution, saturation, and preheating of hydrogen-rich syngas and the energy losses of CO shift, CO_2, and sulfur removal are in the same range as the previously described SFG concept. Hence, the total efficiency is slightly higher and results in about 35.9% due to high temperature heat recovery from the raw gas (figure 12–20). But compared to the previous concept the Shell based IGCC design leads to higher investment and higher plant integration. The water steam cycle represents a three-pressure reheat design and is highly integrated between gasification and gas cleaning. Thus the heat exchanger surface of gas turbine heat recovery generator is basically designed for syngas operation leading to different operational concepts between syngas/hydrogen and secondary fuel operation.

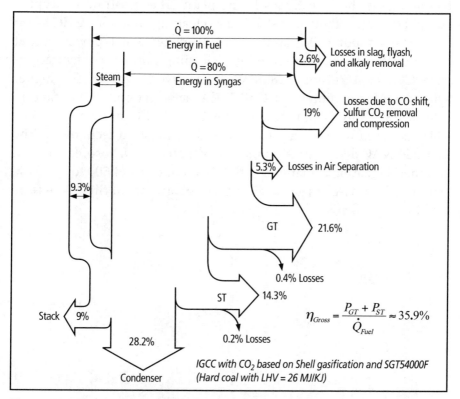

Figure 12–20 Sankey diagram of Shell-based IGCC concept with CO_2 capture

Other IGCC concepts with and without CO_2 capture

Apart from Shell, SFG, and HTW concepts with Siemens combined-cycle technology, other IGCC concepts based on GE energy or MHI gas turbine technology as well as CoP, GE and MHI gasification technology are well known. The IGCC concept of GE energy is primarily based on GE's gasification technology, which is currently in operation in Polk County, Florida, and consists of a coal slurry gasification process with radiant heat recovery for first syngas cooling. For gas cleaning a sulfinol washing unit and GE's advanced F-class gas turbine technology is considered. In case of configuration without CO_2 capture, the plant produces a net power output of about 640 MW; if the plant is designed for 90% CO_2 capture, about 555 MW can be achieved when a two-train concept is assumed.

Another well-known IGCC concept based on F class gas turbine technology is the CoP E-Gas IGCC plant. The two-train concept is equipped with either an MDEA gas cleaning unit for IGCC plant concepts without CO_2 capture that produces a net power output of 623 MW, or a sulfinol-based gas cleaning unit for configurations with CO_2 capture and 518 MW, respectively. Due to the higher syngas methane content of CoP E-Gas gasification, a CO_2 capture rate of 88% only is practicable. The third IGCC concept available today is based on MHI gas turbine and gasification technology. The first 250 MW demonstration plant is situated in Nakoso, Japan, and commercialization is planned with IGCC concepts of 500 to 650 MW sizes based on G-class gas turbine technology and an MHI two-stage air-blown gasifier.

Summary and Investment Cost of IGCC Concepts

Current IGCC concepts consist of different gasifier, gas cleaning, and combined cycle technologies, and their competitiveness is very much depending on feedstock and CO_2 capture rate. Under predefined condition the four selected IGCC concepts represent optimal solutions. For lignite application with CO_2 capture SFG and Shell based IGCC solutions are suitable, too. Alternative IGCC solutions such as GE and CoP E-Gas™-based IGCC application are preferred for high-rank coals and not optimal for low-rank coals because of their slurry feeding limitations.

Today's investment cost calculation for IGCC application are based on front-end engineering design (FEED) studies, which leads to uncertainties of about ±20%. IGCC still presents a new power generation technology with only few demonstration plants in operation. Thus cost basis for components, units, and contingencies are tentative and estimated on database available 2006. Moreover, the price increase of about 30% to 40% due to economy growth between 2005 and 2007 resulted in higher cost uncertainties, and delayed a number of investment decisions.

It can be concluded that a number of IGCC concepts available today vary with regard to performance and investment costs, depending on selected technology, coal quality, and emission requirements; 12% to 30% (with CO_2 capture) of gross power output is internally consumed for syngas generation and gas compression. Therefore, efficiency numbers of IGCC application are generally not applicable, depending on selected technology, component design, and integration features. Although natural gas compression cannot be neglected as well for natural-gas-fired combined-cycle plants, it is typically not considered for efficiency calculation.

IGCC in total suffers on high investment but also offers a high potential for cost reduction and further efficiency improvement when the latest features of gas turbine, air separation, and gas cleaning developments are introduced. IGCC applications are promising for future fossil power generation, focusing on low emission and capability for CO_2 capture.

13 Carbon Dioxide Capture and Storage

After decades of expert and public-level discussion, the existence of global warming and its reasons are undisputed to a far extent today. Increasing ambient temperatures, more frequent heat waves, and disastrous storms and floods are indications. Among all natural or technical gases, which are assessed to be responsible for the global warming effect, CO_2 is the most important one. This results, to a lesser extent, from its physical characteristics, but more from the high amount emitted from all carbon conversion processes. The world population has grown more than four times since the beginning of the 20th century, resulting in steeply rising energy consumption (figure 13–1). In addition, the per-capita energy demand increased as a consequence of the growing prosperity in the industrialized countries. As most of the energy was and is based on fossil (carbon-containing) fuels, the atmospheric carbon dioxide (CO_2) concentration accumulated from 280 up to some 370 ppmv today. In parallel, the global average temperature rose by about 0.8K. According to the Intergovernmental Panel of Climate Change (IPCC) fourth assessment report, a doubling of the CO_2 concentration results in an increase of the global mean temperature by 2 to 4.5K, with a best estimate of 3K.

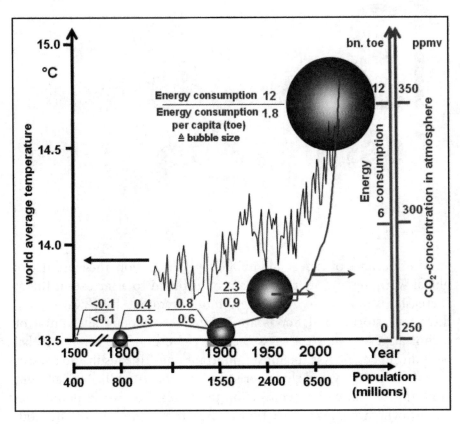

Figure 13–1 Historical trend of energy consumption and global warming

To meet the challenging targets of significant CO_2-emissions reduction a diversified package of measures is needed.

Methods for CO_2-Reduction

In addition to measures, which are aiming at directly reducing the CO_2 concentration in the atmosphere such as biological sinks (photosynthesis, reforestation), there are two fundamental pathways for mitigating the CO_2 problem.

- Efficient production and use of energy
- Substitution of high-carbon fossil fuels by low-carbon/carbon-free fuels

The combination of the two thermodynamical processes—Joule (Bryton) and Clausius-Rankine cycle (combined cycle)—results in high efficiencies and consequently low CO_2, nitrogen, and sulphur oxides emissions. There is a direct correlation among power plant efficiency, saving of primary energy, and emission of CO_2 in case of fossil-fuel-fired technologies (figures 13–2 and 13–3). Most of the combined-cycle plants in operation today are based on natural gas and consist mainly of methane, the fossil fuel with the highest hydrogen to carbon ratio.

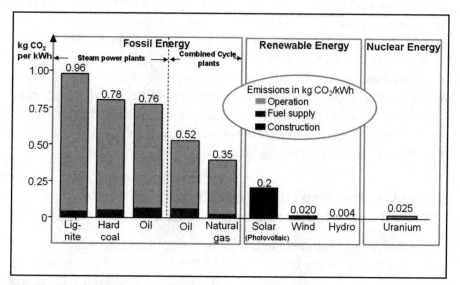

Figure 13–2 CO_2-emissions mitigating by fuel switch

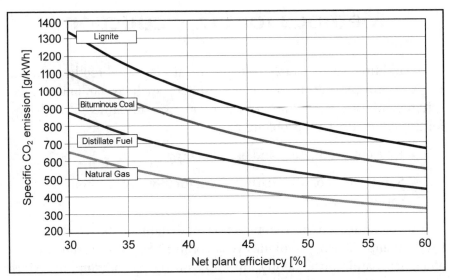

Figure 13–3 Typical specific CO_2-emissions from fossil fuels

Both effects, high efficiency and use of low-carbon fuel, are of advantage with a view to global warming mitigation. On the other side, at least the conventional resources of natural gas are clearly limited and a significant and lopsided switch towards natural gas might be contrary to a safe, independent, and sustained energy supply for many countries worldwide. For this reason, switching to less carbon-intensive fossil fuels doesn't have the potential to significantly reduce the emission of greenhouse gases. In fact, power generation from coal, which is widespread worldwide and available for long-term, has to be taken in an environmental benign way. Beside the development of high-efficient steam-power plants also the combined-cycle technology is to be applied for coal (e.g., Integrated Gasification Combined Cycle, or IGCC).

Reducing the emissions of CO_2 to the atmosphere—carbon capture and storage (CCS)

For that, the CO_2 molecule is in-situ captured and stored. Storage in this connection means the isolation of the CO_2 from the atmosphere for a long time. Carbon capture and storage (CCS) could allow the continued use of fossil fuels, also coal, whereas other CO_2-free energy sources are developed and applied. However, CCS reduces the efficiency of power generation, and, consequently, such measures need more primary energy. In addition, investment and operational costs rise.

CCS is considered first for large-point CO_2 sources, central power generation plants, and large CO_2 emitting industries such as refineries or chemical plants. In principle, the separated CO_2 can be reused as feedstock for synthesizing chemicals or fuels, possibly using CO_2-free produced hydrogen. But these applications are not expected to contribute to a significant abatement of CO_2-emissions. Main options for CO_2-storage are geological formations, selected areas of the worldwide oceans, and depleted oil and gas fields. An application, which is already a commercial option, is to use the CO_2 as driving medium in the enhanced oil recovery (EOR). By injecting CO_2 the pressure of the oil field is enhanced, and density and viscosity of the oil is reduced. Thus, EOR can potentially increase the total recovered production from an oil reservoir by 10 to 30%. In a similar way CO_2 could be used for enhanced gas and coal-bed methane production.

The concepts for CO_2 separation are divided in three fundamental categories.

- Capture of the carbon fraction before being combusted (precombustion capture)
- Capture of the combustion product CO_2 (post-combustion capture)
- Combustion of the fuel with nearly pure O_2 to get CO_2 enriched flue gas (Oxyfuel fired concepts)

Separation of CO_2 from Gas Mixtures

There are various applications for CO_2 removal in the chemical industry or the long-term purification of natural gas. However, the requirements for CO_2 capture processes when being implemented into power generation concepts are partly different and more challenging: The behaviour of the power plant caused by the grid demand is very dynamic, and the volume flows to be handled are higher. In addition, oxygen and trace elements such as sulphur or alkali compounds may cause degradation of the solvent.

Sorption

The CO_2 is loaded to a liquid (absorbent) or solid (adsorbent), and released under changed temperature and/or pressure conditions (desorption). With view to the reversible bonding mechanism it is differentiated between physical and chemical sorption with a smooth transition. Physical absorption depends strongly on temperature, and can be used in a very effective way at increased total gas pressures and high content of the compounds to be removed. Due to Henry's law, which is exactly valid for diluted solutions only, the amount of removed substance is nearly proportional to its partial pressure (figure 13–4). As a rule chemical absorbents are more efficient at low gas pressures.

Hereby, an aqueous alkaline solvent, usually an amine, reversibly reacts with the acid CO_2 gas such as for monoethanolamine (MEA):

$$2RNH_2 + CO_2 \Leftrightarrow RNH_3^+ + RNHCOO^- \text{ (R: organicgroup)} \quad \textbf{(13–1)}$$

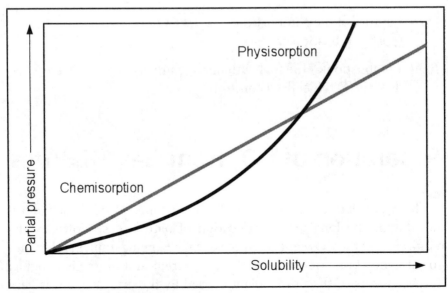

Figure 13–4 Typical trend of solubility in physical and chemical washing agents

Besides cycling the solvent, the main energy consumption is caused by the desorption unit. Depending on capture rate, desorption temperature (reboiler temperature, see figure 13–11) and solvent flow rate, the steam consumption such as for the MEA process was found equal to around 4 MJ/kg CO_2. As a consequence of degradation and loss in activity, some makeup is always needed, which contributes to the operating costs. Particularly for CO_2 from flue gases (post-combustion), new or modified solvents are under investigation for high CO_2 loading, high lifetime, and low energy demand.

Membranes

Membranes for gas separation are thin layers made from porous materials, which are selectively permeable for gases. Plant concepts based on membranes are considered to be less complex and more cost-effective. CO_2-selective membranes are already applied in industry, such as for the treatment of natural gas, but have to be further developed for large-scale plants, high stability, and availability, as well as the application at high temperatures.

Generation of Oxygen

Precombustion and Oxyfuel fired CCS concepts use highly oxygen-enriched air, which results in lower dilution and, thus, improved reaction conditions as well as reduced dimensions of apparatus and pipings (lower investment). On the other side an additional process unit for air separation is needed, which makes the overall plant more complex and increases the internal power demand.

Cryogenic temperature air separation unit

Today, large-scale air separation necessary to provide precombustion or Oxyfuel plants in the order of hundreds of megawatts are based on proven air distillation at cryogenic temperatures. For reduced power demand the product oxygen fraction may be pumped instead of being compressed. Single-train plants are available for more than 4,000 t/d with typical 95% to beyond 99% purity of the oxygen fraction.

Ion transport membranes

For these innovative concepts the oxygen is transported as an ion through a membrane of adapted lattice structure (figure 13–5). These ion transport membranes (ITM, or specifically OTM in case of oxygen) are highly selective. The active thin layer is manufactured from ceramic material and applied on a porous carrier for mechanical stability. For sufficient separation capacity the ionic mechanism needs high operating temperatures of about 850°C, which can be realized, for example, by directly heating up the air through combustion.

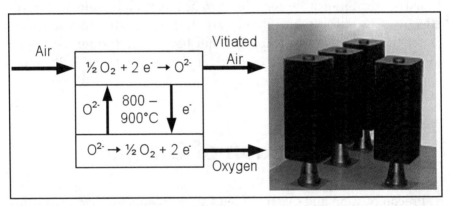

Figure 13–5 Principle of ion transport membrane

Chemical looping

In the chemical looping concept there is no direct contact of combustion air and fuel. In fact, the oxygen is fixed on a solid carrier medium and then moved to another, separate area where it is used for oxidizing. The reduced carrier is cycled back to the air to be loaded again (figure 13–6). As a consequence, no NO_x formations can occur. Typical materials are metals such as iron, nickel, copper and manganese. The process is currently in the state of a pilot plant.

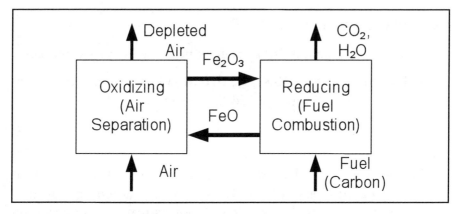

Figure 13–6 Principle of chemical looping combustion

Precombustion Capture

The feedstock is processed in a way to get the carbon content ready for separation. This is realized by reaction of the carbon to CO, which is then converted to CO_2 and hydrogen (figure 13–7).

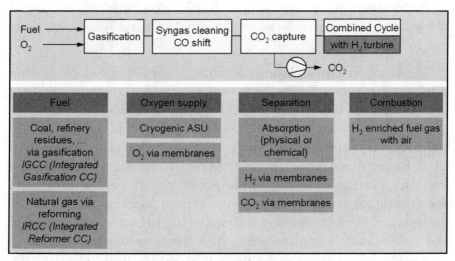

Figure 13–7 Schema of precombustion CO_2 capture

Conversion of carbon to CO

There are two reaction paths resulting in the synthesis gas (syngas).

Partial oxidation (POX), gasification. Reaction with oxygen; typically 1250 to 1500°C (entrained-flow gasification).

$$C_xH_y + x/2O_2 \Leftrightarrow xCO + y/2H_2, \text{ exothermic} \qquad (13\text{-}2)$$

The syngas is cooled by quenching (water, recycled gas) and/or through a heat recovery system. By using nearly pure oxygen instead of air, the volumes to be handled at downstream CO_2 separation are reduced.

Steam reforming. Reaction with H_2O; typically 800 to 900°C.

$$C_xH_y + xH_2O \Leftrightarrow xCO + (x + y/2)H_2, \text{ endothermic} \qquad (13\text{-}3)$$

The process is catalysed and indirectly heated. A waste heat system is used for cooling down the products to the temperature needed for the following shift reaction. Steam reforming followed by pressure-swing absorption is today's way to produce highly pure hydrogen from natural gas.

Autothermal reforming (ATR). Combination of partial oxidation and steam reforming (figures 13–2 and 13–3); typically 950 to 1050°C.

The reaction takes place in a refractory-lined vessel containing a nickel catalyst bed. The heat needed for the reforming process is directly produced from partially oxidizing the feedstock. There is no indirect heating by combustion, which would need a second separate CO_2 capture plant when aiming for high capture rates in total. In addition, compared to steam reforming, the process is more compact, favorable for large-scale plants, and the investment costs for the reactor are lower. To prevent carbon soot deposition on the catalyst and to limit methane slip, steam has to be added in excess of the stoichiometric requirements. ATR needs an oxygen generation plant, which also provides nitrogen, if necessary, for moderating the combustion conditions.

Conversion of CO to CO_2 (CO shift). The temperature level for this process depends on the applied catalyst (180 to 500 C). Its sensibility to sulphur compounds needs the shift to be performed up- (raw gas shift) or downstream (clean gas shift) of the desulphurisation unit. To cope with the excess heat of the reaction, a two-stage adiabatic reactor system with intermediate cooling is applied.

$$CO + H_2O \Leftrightarrow CO_2 + H_2, \text{exothermic} \tag{13-4}$$

Separation of the CO_2

Gasification and downstream combustion of the syngas in a gas turbine take place under elevated pressure levels. Compared to post-combustion capture, the CO_2 partial pressure is higher in the syngas. Thus, the CO_2 separation unit is smaller in size and the specific energy consumption is lower. Physical solvents such as methanol (Rectisol process) are most suitable to such conditions. Depending on the applied shift systems and the requirements for transport and storage, both sulphur compounds and CO_2 are washed out together or selectively.

Use of the remaining H_2 enriched gas for power generation

Independent of the feedstock, also for coal as a solid or liquid hydrocarbon, the precombustion decarbonization process ends with a hydrogen-enriched fuel in gaseous state usable in a gas turbine. This is why precombustion concepts are predestined to be integrated into a combined-cycle process:

In the case of natural gas, the concepts are known as Integrated Reforming Combined Cycle (IRCC) (figure 13–8). For coal or refinery residues, they are called Zero Emission Integrated Gasification Combined Cycle (ZEIGCC). The basic IGCC process is described in Chapter 12. In principle, the modifications needed for carbon capture are limited to the gas treatment and the gas turbine combustion system. But as a consequence mainly of the high steam demand for the CO shift, the interactions between the water/steam cycle and the gas island are also affected. By reducing this steam export, gasification systems applying a water quench system for cooling the raw gas downstream of the gasifier may be advantageous for ZEIGCC concepts.

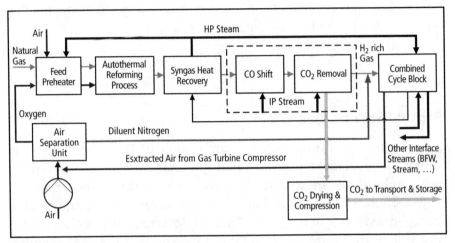

Figure 13-8 Combined autothermal reforming and CO_2-capture

Table 13-1 shows typical compositions of the decarbonized, undiluted synthesis gas from the CO_2 removal for different processes and feedstocks. For the use in the combustion systems of modern high-temperature gas turbines, such hydrogen-enriched gases have to be conditioned.

Table 13-1 Typical hydrogen enriched syngases downstream gas cleaning

		IRCC Natural Gas / ATR	ZEIGCC Hard Coal / SCGP	ZEIGCC Lignite / HTW
H_2	vol%	95.2	85.3	81.7
CO	vol%	0.7	4.8	0.8
CO_2	vol%	0.5	0.5	0.1
CH_4	vol%	1.6		6.6
N_2 + Ar	vol%	1.4	9.4	10.9
H_2O	vol%	0.6		
LHV	MJ/kg	77.2	36.4	41.4

The different combustion behavior of hydrogen compared to the standard fuel natural gas results in special requirements for burner design and gas turbine operation (Table 13–2).

Table 13–2 Characteristics of typical gas turbine fuels

		CH_4	H_2	CO
LHV	MJ/kg	50.0	119.9	10.1
	MJ/m³STP	35.8	10.2	12.6
Flame speed in air	cm/s	43	350	20
Stoichiometric combustion temperature	K	2227	2370	2374
Density	kg/m³STP	0.74	0.09	1.25
Specific heat	kJ/kg K	2.18	14.24	1.05
Flammability limits	vol%	5–15	4–75	12.5–74

Important for the geometric burner design is the substantially smaller density leading to higher volumetric flow rates. The use of today's natural gas burners for hydrogen-rich gases would lead to high fuel injection velocities and high fuel-side pressure losses. Thus, the burner nozzles need to be redesigned. An additional crucial characteristic is the very high reactivity and laminar flame speed of hydrogen. Special burner modifications are necessary to avoid the risk of flame flashbacks and pre-ignition. These modifications ensure stable combustion at low NO_x formation. For this, a homogeneous fuel/air mixture and the avoidance of recirculation areas within the burner are of particular importance. These are the specifications for the development and optimization of premix burner concepts, applied for innovative high-temperature gas turbines. The design shown in figure 13–9 is derived from the Siemens standard natural gas premix burner. A concentrical passage is added for enabling the high volumetric throughput of the hydrogen-rich gas, feeding directly downstream to the diagonal swirler into the fuel/air reaction zone.

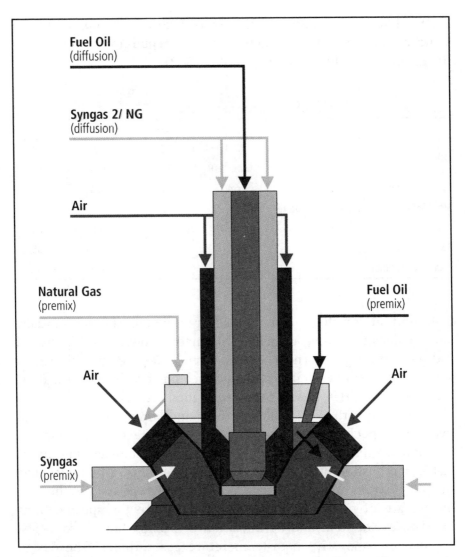

Figure 13–9 Concept of an innovative premix burner for hydrogen-rich syngases

The development of gas turbines for decarbonized fuels is pushed by all manufacturers and supported by different national and international funding programs. One of the first large research and development programs initiated by the European Commission was the Enhanced CO_2 Capture (ENCAP) project, which started in 2004 and put together a great number of partners from research institutes, universities, manufacturers, and utilities.

Capture of the CO₂ after Combustion (Post-combustion)

The carbon content reacts to CO_2 through air-blown, hyperstoichiometric combustion (with possible traces of CO), which is separated for storage. The remaining flue gas, which mainly consists of nitrogen and H_2O, is discharged to the atmosphere (figure 13–10).

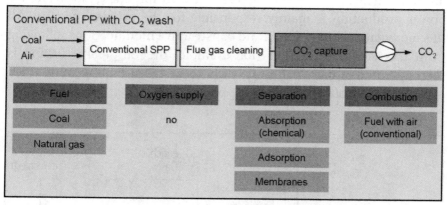

Figure 13–10 Schema of post-combustion CO_2-capture

Post-combustion capture methods can be applied to all power generation concepts because they are located in the flue gas stream at the end of the process chain. Thus, this capture method is best qualified for being incorporated into conventional coal-fired steam power plant concepts, for which no other CO_2 capture paths are available. In principle, also retrofitting of existing plants is an option for post-combustion capture. In case of natural-gas-fired combined-cycle plants, post-combustion capture has to compete with the two other capture methods—IRCC and Oxyfuel firing.

The exhaust gases to be treated are at atmospheric pressure, of high volume flow, and relative low in CO_2 content, particularly in the case of a natural-gas-fired combined-cycle plant. For CO_2 separation under such conditions, absorption processes applying mainly chemical sorbents are preferred today. Figure 13–11 shows the flow sheet of such a sorption cycle. The CO_2 is transferred to the counter-flowing solvent in the absorption column, which leaves the column at its bottom and is pumped to the top of the desorber unit. For regenerating the chemical solvent, such as breaking its bonds with CO_2, heat at 100 to 140°C is needed, which is extracted from the water steam/cycle to the desorption unit. The medium for heat transfer is steam at the corresponding temperature level for desorption. In the current concepts, about half to two-thirds of the LP steam are extracted from the IP to LP crossover pipe. The condensate is fed back to the water/steam cycle downstream of the condenser. The resulting loss in expansion energy available for power production is mainly responsible for the process' penalty. All thermodynamic, operational, and constructive modifications following from this steam extraction (for example, the LP steam turbine is then operated in part load only) have to be taken into account when considering retrofitting an existing power plant.

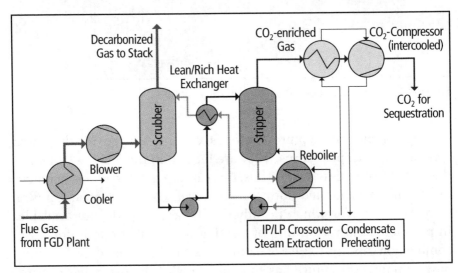

Figure 13–11 Post-combustion absorption process

Beside the conditions previously mentioned and, of course, the specific load rate and energy needed for desorption, a criterion for solvent selection is the potential degradation that is caused, for example, by oxygen and possible corrosiveness. Although various chemisorption agents and processes are already used in the chemical and other industries, predominantly aqueous solvents of amines (table 13–3), the different requirements need further development and demonstration.

Table 13–3 Commercially available CO_2 absorption systems for post-combustion applications

Kerr-McGee/ABB Lummus Crest Process (US)	MEA (aqueous 15-20 wt %) + inhibitor
Fluor Daniel ECONAMINE™ Process	MEA (aqueous 30 wt %), with inhibitor
Kansai Electric Power Co., Mitsubishi Heavy Industries, Ltd.	Sterically-hintered amine (KS solvents)

For example, in the large European research and development project CO_2 capture and storage (CASTOR), several solvents are tested for CO_2 removal in a flue gas side-stream of the Elsam coal-fired power station in Esbjerg/Denmark. Against the background of obligations for CO_2 reduction and the increasing application of EOR, several projects are pushed for developing CO_2 capture from natural-gas-fired combined cycles mainly in Norway, such as Karstoe and Tjeldbergodden (Statoil).

The SARGAS cycle (a concept belonging to the Norwegian-based company SARGAS AS) describes a post-combustion concept where the CO_2 is advantageously separated under elevated pressure (figure 13–12). It is a combined cycle applying an open gas turbine, which is externally fired in a pressurized boiler. The flue gas is cooled, decarbonized, and cleaned from NO_x. After being heated up again, it is expanded through the turbine unit. About 15 to 20% of the total power is generated by the gas turbine. The rest is generated in the steam cycle, which is heated by the combustion chamber.

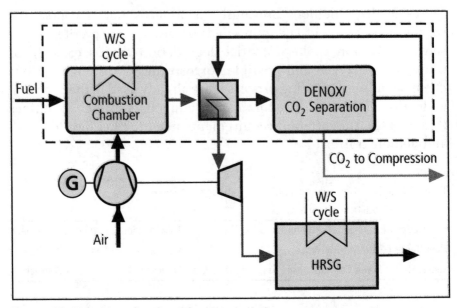

Figure 13–12 Post combustion CO_2 capture under pressure

Oxyfuel Processes

From the basic idea, Oxyfuel concepts are post-combustion processes, as well. However, by using nearly pure oxygen for the combustion, the flue gas is enriched in CO_2 and consists to a great extent of CO_2 and H_2O, which is condensed to get the CO_2 for deposing. Thus, strictly speaking, it is not the CO_2 that is separated. As there are modifications and additional steps needed up- and downstream the combustion, these Oxyfuel concepts are also called integrated CO_2 capture (figure 13–13).

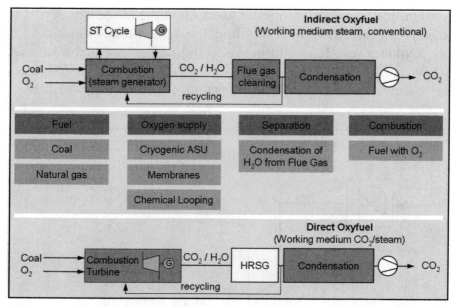

Figure 13–13 Schema of Oxyfuel firing CO_2 capture

For reducing the combustion temperature, which for pure oxygen is about 3,500°C, down to the level of conventional boiler or gas turbine conditions, either parts of CO_2 have to be recycled or water has to be injected.

With view to the medium used in the working machine, two systems are distinguished: indirect Oxyfuel systems where heat is generated in an oxygen-blown combustion and indirectly transferred to the working medium, and direct Oxyfuel systems, where the CO_2/H_2O flue gas is directly used in a turbine.

There are several Oxyfuel-fired concepts proposed or under development. Some of these are discussed in the following sections.

In the Graz cycle (Graz University of Technology) the fuel is burnt with the stoichiometric mass flow of oxygen, which is produced externally (over the fence), such as in a cryogenic air separation unit (figure 13–14). Steam and recycled CO_2 are used for cooling purposes. The flue gas is expanded in two steps with an intermediate heat recovery system, and is divided into water and CO_2 in the condenser. A part is then recompressed to the combustor. The water is used as medium for the Rankine cycle. The gas turbine working medium is a mixture of CO_2 and steam. It has to be essentially redeveloped because its characteristics are clearly different from the working fluid of a standard gas turbine.

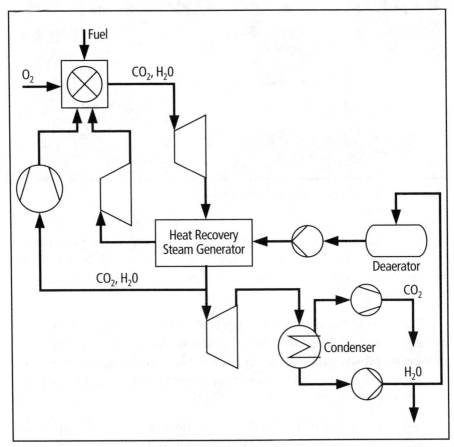

Figure 13–14 Concept of the Graz cycle

The CES cycle (Clean Energy Systems Inc.) uses a modified rocket burner, which is fired by a gaseous fuel and oxygen (figure 13–15). Water that has been separated from the flue gas is recycled for cooling this gas generator. The hot exhaust gas is directly expanded through a multistage turbine to generate power (direct Oxyfuel), or alternatively can generate steam in a heat recovery steam generator to be used in a water/steam cycle (indirect Oxyfuel). H₂O is condensed after the cooling step.

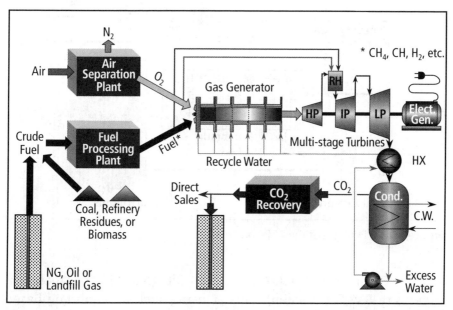

Figure 13–15 Direct Oxyfuel fired CO_2 capture, CES cycle

The Advanced Zero Emission Power plant (AZEP) (Norsk Hydro) combines two nonstandard gas turbines (figure 13–16). The first one is a hot air turbine in principle, where the air is oxygen-depleted by an ion transport membrane and indirectly heated by Oxyfuel combustion (indirect Oxyfuel). This turbine is combined with a water/steam cycle. The combustion products, CO_2 and steam, are used to drive the second turbine, which is operated in simply cycle mode (direct Oxyfuel). CO_2 is separated by condensing the water content of the flue gas. The separation process is supported by continuously reducing the oxygen concentration on the product side by the reaction and by continuously diluting with CO_2/H_2O (sweep gas).

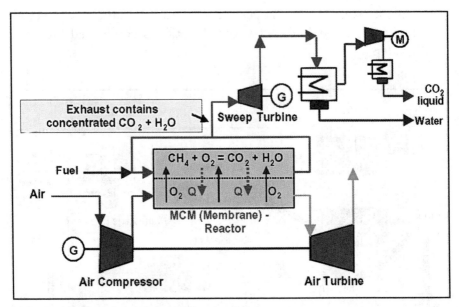

Figure 13–16 Principle of the AZEP process

The German OXYCOAL-AC project (led by the technical university RWTH Aachen) aims at using such a membrane-based Oxyfuel concept for coal-dust-fired power plants.

A promising, long-term option is the zero emission solid oxide fuel cell (ZESOFC) concept. It combines an air-blown gas turbine with the high-efficient fuel cell process (figure 13–17). The solid oxide fuel cell (SOFC) has to be modified for combustion of the anode exhaust gas with oxygen. That way, the completely burnt flue gas consists of CO_2 and H_2O only, which is expanded in a specially designed turbine.

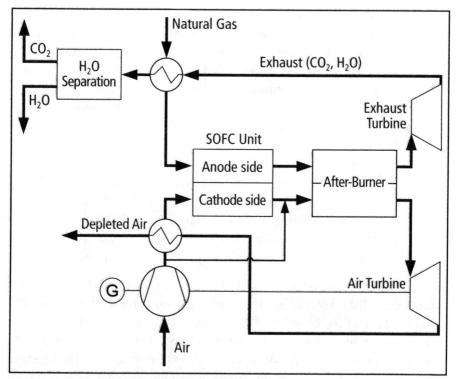

Figure 13–17 Principle of the ZESOFC process

Impact on Efficiency and Economy

Because of the additional process units and power demand, CO_2 capture leads to reduced power generation efficiency and increased costs. Various studies are made to quantify these impacts for the different CCS concepts resulting in a large bandwidth because of the different boundary conditions and steps of development, or uncertainties when designing novel processes in general. In fact, certain information about plant performance and costs will be available from the ongoing first demonstration projects only.

Based on public research and development as well as on utility and manufacturer in-house studies, a comprehensive technology benchmark has been done by the ZEP Working Group 1 (European technology platform for zero emission fossil fuel power plants) encompassing technologies, which are expected to be in use in large applications by

2020. By an efficiency penalty of about 10 percentage points (reference 58%) combined-cycle concepts on natural gas are in the same range as hard-coal-fired power plants. No difference regarding efficiency results for the three principle paths pre- and postcombustion and Oxyfuel. However, the latter are assumed to lead to a 100% CO_2 capture rate (natural gas), whereas for post-combustion 85% and for precombustion 93% are the bases. With view to the CO_2 avoiding costs, CCS makes sense first of all for base-load operated coal plants (figure 13–18 top) for two reasons. For coal fired plants the specific CO_2 concentrations are higher and, as a consequence of the lower fuel costs, the reduction in efficiency for the CCS case results in a lower impact on the additional costs. On the other side, the amount of CO_2 produced per MWh electricity is lower for gas-fired plants. As a consequence the tendency levels off when focusing on the power generation costs (figure 13–18 bottom).

A comparison focusing on various natural-gas-fired gas turbine cycles has been published by the Norwegian independent research organization SINTEF. From that, based on 57% net plant efficiency for the reference case the near term pre- (air-blown ATR) and post-combustion (MEA) concepts lead to a drop down to 47% and 48%, respectively. In the same range are Oxyfuel processes applying a cryogenic air separation unit with the potential up to 49% for an advanced version of the Graz cycle (S-Graz). Higher efficiencies between 50 to 53% are evaluated for Oxyfuel processes using future membranes for the oxygen production or chemical looping combustion. The study also included the highly advanced combined SOFC/gas turbine cycle with an efficiency potential of 67%, which, however, needs clearly long-term development for large-scale applications.

In addition to thermodynamics and the cost situation, further aspects are to be considered when evaluating CCS technologies, such as the quality of the CO_2 stream with view to its application/storage, the plant size, the complexity of the concept that might significantly effect the availability, or the suitability for retrofitting to existing plants. And, of course, it is not only the capture technology that influences the economy of CCS. For example, the absolute fuel price has a significant influence on the CO_2 avoidance costs. Therefore, according to European Technology Platform for Zero Emission Fossil Fuel Power Plants (ETP-ZEP) the decision on future capture plants will be driven primarily by the fuel choice and only second by the capture technology.

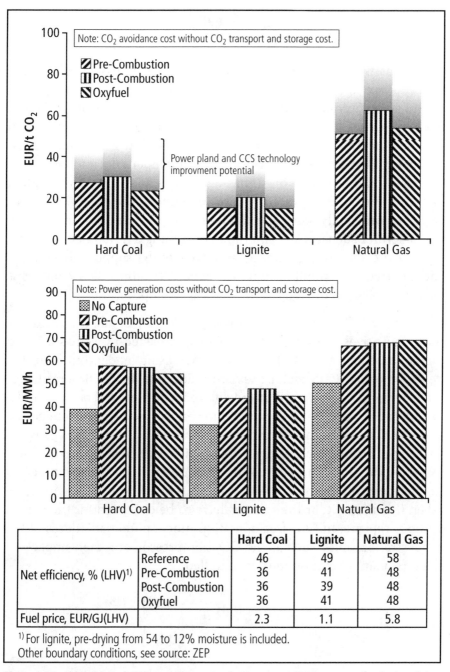

Figure 13–18 Expected CO_2 avoidance and power generation costs for industrial-scale power plants in operation by 2020 (Source: ZEP)

Outlook

For gas-fired combined-cycle plants a diversity of CO_2 capture concepts are under progress particularly for Oxyfuel combustion. In case of direct Oxyfuel the working medium, a mixture of mainly CO_2 and H_2O, significantly differs from the conventional flue gas from air-blown combustion. This means that the specific mass flow and compression/expansion work as well as the turbine outlet temperature and the design of blades and dimensions of gas passages. Thus, for the breakthrough of direct Oxyfuel in this sector the core component, the gas turboset, has to be fundamentally redesigned and optimally adapted. For the turbine manufacturer such a costly and time-consuming development requires a positive economical perspective as well as the support by national and international funding organizations. Altogether, this concept is expected to be a more long-term CO_2 capture option.

To apply CCS measures in power generation beside a technology push, incentives from market or policy are needed as well as a broad public acceptance of CO_2 storage. Increasing endeavours on the development of CCS techniques are made worldwide. Several large-scale demonstration plants for the different concepts will be built in the next decade (table 13–4). For future power plant projects the obligation is under discussion to include CO_2 capture or at least to be ready for CCS; this means, for example, to provide sufficient space for later installing the additional systems or to consider future steam extraction for the desorption units. Moreover, the reduction in power must be taken into account, or the systems have to be oversized for the period without capture. In CO_2-free operation mode, potential failures of the sequestration units, such as the CO_2 compressor, and their impact on the total plant operation must be handled.

Table 13–4 Selection of European CCS demonstration projects (Source: The World Energy Book, issue 3)

Country	Start-up	Owner	Capture	Storage
Denmark	2005	Castor project, FP6	Post-combustion; experimental; coal-fired plant	A few '000 t/y; no storage
Germany	2008	Vattenfall	Oxyfuel boiler; coal-fired plant; 30 MW	Pilot scale; no storage
France	2008	Total	Oxyfuel boiler; 30 MW, gases and liquids	Pilot scale; Lacq plant
Norway	2009	Stratkraft	Gas-fired power plant Karstoe, 400 MW	EOR (Field not specified)
UK	2009	Progressive Energy	Pre-combustion 800 MW; IGCC	—
Netherlands	2008	SEQ International, ONS Energy	Oxyfuel; gas-fired plant; 350 MW	EOR/EGR Drahten
UK	2010	BP/Scottich and Southern Energy	Pre-combustion; gas-fired plant; 350 MW	EOR Miller field
UK	2010	Powerfuel	Pre-combustion; IGCC; 900 MW	—
Norway	2011	Shell/Statoil	Gas-fired power plant; 860 MW	EOR Draugen/Heidrun oilfields
UK	2011	E.On	Pre-combustion; IGCC	Offshore
Germany	2014	RWE	Pre-combustion; IGCC; 450 MW	Geological storage
UK	2016	RWE	Coal-fired power plant retrofit	To be determined

14 Typical Combined-Cycle Plants

Table 14–1 shows how the technologies described in previous chapters are applied from the different suppliers in the global electric power industry.

Table 14–1 Combined-Cycle Plant examples—overview

Plant Name	Country	Supplier	Net-Output MW	Configuration	Remark
Taranaki	NZ	ABB*	360	single-shaft	
Monterrey	MEX	ABB*	2×242	2 single-shaft	
Phu My 3	VN	Siemens	716	multishaft	
Palos de la Frontera	E	Siemens	3×380	3 single-shaft	
Arcos III	E	GE	823	multishaft	
Diemen	NL	ABB*	250	multishaft	Cogeneration
Shuweihat IWPP	UAE	Siemens	1495	multishaft	Desalination
Vado Ligure	IT	Ansaldo	780	multishaft	Repowering
Puertollano	E	Siemens	300	multishaft	IGCC
Monthel	CH	KAM/Siemens	55	multishaft	Cogeneration

*ABB (today ALSTOM)

The Taranaki Combined-Cycle Plant, New Zealand

The 360-MW Taranaki combined-cycle power plant (figure 14–1) is owned by Stratford Power, Ltd., New Zealand's first independent power producer, and supplies electricity as a baseload facility connected to the North Island transmission grid.

Figure 14–1 View of Taranaki combined-cycle plant

The plant is located about 50 kilometers from New Plymouth on a site 270 meters above sea level. Initially the plant will fire only natural gas taken from the Maui gas field, but the design enables conversion to fire oil in the future, should this prove necessary.

The power plant consists of a single-shaft block with one ABB GT26 gas turbine, one ABB natural circulation HRSG, and one ABB two-casing steam turbine, as well as the necessary auxiliary equipment. Due to the lack of cooling water on the site a wet cooling tower has been installed.

A high-efficiency triple-pressure reheat cycle was chosen for the plant, with HP live-steam at 103 bar (1479 psig)/568°C (1,054°F), reheat live steam at 24 bar (333 psig)/568°C (1054°F), and LP steam at 4.0 bar (44 psig)/saturated. The gas is preheated to further increase the efficiency (figure 14–2).

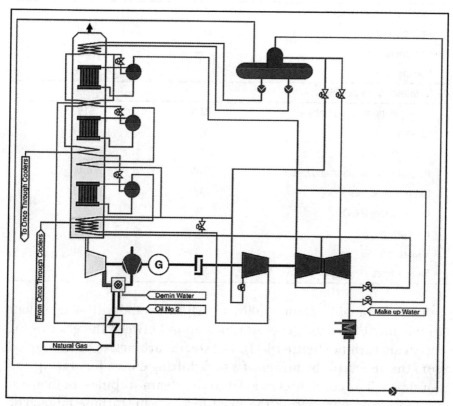

Figure 14–2 Process diagram of Taranaki combined-cycle power plant

Table 14–2 summarizes the performance data of the Taranaki combined-cycle power plant.

Table 14–2 Main technical data of Taranaki combined cycle power plant

Plant Configuration:		KA 26-1
Number of blocks		1
Number of gas turbines/block		1
Main fuel		Natural gas
Ambient Conditions:		
Ambient temperature	°C / °F	11.6 / 53
Ambient pressure	mbar	981
Relative humidity	%	84
Cooling type		Cooling tower
Performance data per block:		
Total fuel input to gas turbines	MW	615
Gross output	MW	359.9
Gross efficiency (LHV)	%	58.5
Auxiliary consumption and losses	MW	6.1
Net power output	MW	353.8
Net efficiency (LHV)	%	57.5
Heat rate (LHV)	kJ/kWh	6261
Heat rate (LHV)	BTU/kWh	5934
Process heat	MJ/s	0

The Taranaki plant follows ABB's standard single-shaft arrangement with the generator located in between the gas turbine and steam turbine (figure 14–3). The steam turbine can be decoupled from the generator by means of a self-shifting clutch for startup and shutdown. The condenser is axial to the steam turbine—both are at ground level—eliminating the need for a steam turbine table. The hydrogen-cooled generator is mounted on a skid, together with its lube oil system and hydrogen coolers. The skid is mounted on transverse rails so it can be moved to the side for rotor inspections. The arrangement is very compact, with a footprint of the combined-cycle block measuring about 38 by 96 meters.

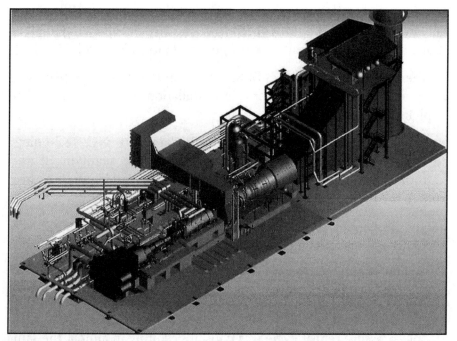

Figure 14–3 Arrangement of Taranaki combined-cycle power plant

This plant is the first in a series of ABB-standard reference single-shaft units for the 50 Hz market. Some of the follow-up orders are in the United Kingdom, Chile, and Japan.

ABB had also the responsibility for the operation and maintenance of the plant over the first six years. An online monitoring system is connecting the plant to the ABB headquarters in Switzerland, where direct assistance will always be available to the plant operators.

The 480 MW Combined-Cycle Power Plant, Monterrey, Mexico

The Monterrey combined-cycle power plant has been ordered by Comisión Federal de Electricidad (CFE), the national utility of Mexico, and achieved commercial operation in the year 2000. The plant, located in Monterrey (Nuevo Leon), is designed for pure electrical power generation with a net output of 484 MW and 56% net efficiency, making it one of the most efficient thermal power plants in Mexico.

The plant is based on ABB's KA24-1 standard reference plant and is among the first units of a series of installations in the 60 Hz market, applying the same design.

The gas turbines are equipped with the latest low NO_x technology to minimize emissions. Effluents are basically nil, qualifying it as a zero-discharge site. An air-cooled condenser uses ambient air as heat sink to the cycle, keeping water consumption to a minimum.

The Monterrey power plant is made up of two identical 242 MW blocks, each comprising one 160 MW ABB GT24 gas turbine, one ABB CE once-through HRSG, and one ABB 90 MW double-casing reheat steam turbine unit. State-of-the-art combined-cycle plants are mostly designed with a triple-pressure reheat cycle to achieve high efficiency, but not so this plant. As explained in chapter 5, a simpler double-pressure reheat cycle was chosen, resulting in almost the same high net efficiency level.

The hot exhaust gases of the GT24 are used to generate steam at 160 bar (2310 psig) for best efficiency. Producing steam at this pressure helps to avoid an IP system, therefore, simplifying the cycle. The once-through HRSG avoids a thick-walled HP drum and results in a high thermal flexibility. The LP part is directly fed from the condenser. Steam of 160 bar (2310 psig) results in small steam volumes that can be efficiently expanded in the geared HP steam turbine. The barrel-type HP steam turbine design provides high thermal flexibility.

This simpler, dual-pressure reheat plant exceeds 56% net efficiency with an air-cooled condenser at 30°C (86°F) ambient temperature. The gas turbine and the steam turbine drive a common air-cooled generator installed between them. Although the gas turbine is rigidly coupled to the generator, the steam turbine is equipped with a self-shifting and synchronizing clutch, enabling startup of the gas turbine independent from the steam turbine. The steam turbine clutch engages automatically during startup as soon as the speed of the steam turbine reaches that of the generator.

The chosen cycle for the Monterrey plant produces HP steam of 160 bar (2310 psig)/565°C (1049°F) and LP steam of 7 bar (87 psig)/320°C (608°F), which are both fed into the steam turbine (figure 14–4). After partial expansion of the HP steam in the HP-steam turbine, it is reheated in the HRSG at 37 bar (522 psig) to 565°C (1049°F).

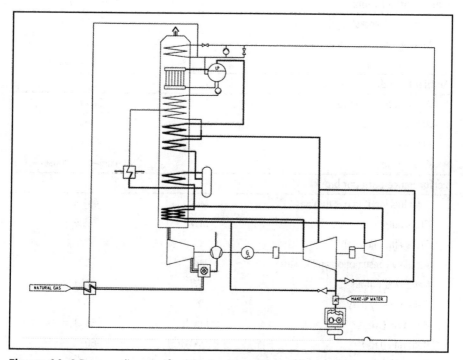

Figure 14–4 Process diagram for Monterrey combined-cycle power plant

The hot exhaust gases are thereby cooled from approximately 650°C (1202°F) to below 100°C (212°F) before being exhausted through the stack. Due to lack of water at this site an air-cooled condenser is used to condense the steam at the outlet of the steam turbine. The auxiliaries are cooled by air blast coolers.

Table 14–3 summarizes the performance data of the Monterrey combined-cycle power plant. As shown in the table, more than 56% of the fuel energy is converted to electrical power, despite the air-cooled condenser operating at 30°C (86°F) ambient temperature that reduces the steam turbine output and increases auxiliary consumption. By this means, plant water consumption is kept to a minimum and no water body is heated up by the dissipated heat. NO$_x$ levels in the exhaust are below 25 ppm.

Table 14–3 Main technical data of Monterrey combined cycle power plant

Plant Configuration:		2 × KA 24-1
Number of blocks		2
Number of gas turbines/block		1
Main fuel		Natural gas
Ambient Conditions:		
Ambient temperature	°C / °F	30 / 86
Ambient pressure	mbar	969
Relative humidity	%	60
Cooling type		Air-cooled condenser
Performance data per block:		
Total fuel input to gas turbines	MW	431.3
Gross output	MW	249.9
Gross efficiency (LHV)	%	57.9
Auxiliary consumption and losses	MW	7.8
Net power output	MW	242.1
Net efficiency (LHV)	%	56.1
Heat rate (LHV)	kJ/kWh	6413
Heat rate (LHV)	BTU/kWh	6078
Process heat	MJ/s	0

Each gas turbine, and the corresponding steam turbine, is arranged indoors in a common machine house, with sufficient laydown area for inspections. The two blocks are fitted in separate buildings. The HRSGs are arranged outdoors. The balance of the plant is arranged to provide short pipe and cable routing to the different balance of plant equipment (figure 14–5).

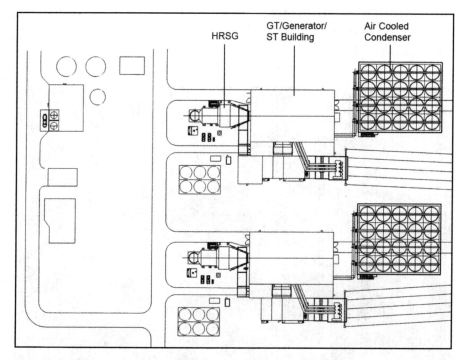

Figure 14–5 Layout of the Monterrey combined-cycle power plant

The 716-MW Combined-Cycle Power Plant Phu My 3, Vietnam

The Phu My 3 combined-cycle plant was built as a build-operate-transfer (BOT) project owned by the PhuMy 3 BOT Power Company. This project-specific company is a consortium of oil and gas major BP, SembCorp Utilities (majority owned by the Singapore government), and the Japanese companies Kyushu Electric and Sojitz (acting as one partner in the consortium). Each of the partners has a one-third interest. A power purchase agreement with Electricity of Vietnam (EVN) is scheduled to run for 20 years, after which ownership will be transferred to the state-owned utility.

Phu My 3 is located 70 km southeast of Ho Chi Minh City near the village of Phu My on an industrial site that already accommodates three plants (figure 14–6).

Figure 14–6 View of the Phu My 3 combined-cycle power plant

In May 2001, the consortium signed an EPC contract with Siemens Power Generation, including a 12-year maintenance agreement. Construction work began in late 2001, and commercial operation started in March 2004.

For achieving best performance at the specified ambient conditions in Southeast Asia, a triple-pressure, single-reheat cycle was selected. The plant rated capacity at ISO conditions is 760 MW with a design efficiency of 58% at the generator terminals.

The multishaft combined-cycle configuration is made up of standardized modules, the main modules being the two gas turbine generator sets, two outdoor heat recovery steam generators, one steam generator set, water-steam cycle system, instrumentation and control equipment, electrical power system as well as pre-designed, pre-engineered buildings or civil structures (figure 14–7).

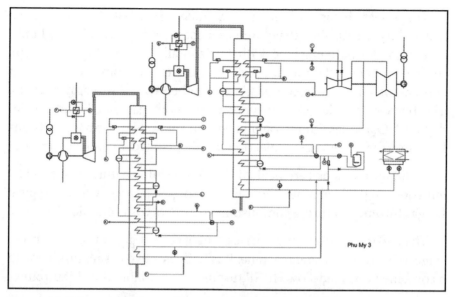

Figure 14–7 Process diagram of the Phu My 3 combined-cycle power plant

The arrangement of the plant is based on the Siemens combined-cycle reference power plant SCC5-4000F 2x1 (figure 14–8).

Figure 14–8 Arrangement of the combined-cycle reference power plant SCC5-4000F 2×1

The selected gas turbine is a Siemens SGT5-4000F machine, a heavy-duty gas turbine, structured as a completely pre-engineered unit. It is a single-shaft machine with horizontally split casings, disk-type rotor with center tie bolt and cold end drive. The combustion system consists of an annular combustion chamber with 24 hybrid burners. The burners are designed to operate on both gas and liquid fuel in dry low NO_x mode without water injection. The system also has provisions for water/fuel oil emulsion injection for NO_x control in diffusion mode.

Pollutant emissions are negligible when burning natural gas, and CO emissions in the load range above 60% are below 10 ppm, which is typical for gas turbines featuring combustion chambers with ceramic tiles.

The steam turbine, which drives an air-cooled generator, is a three-stage, reheat, two-casing machine. It is built as a two casing turbine with a combined opposed-flow HP/IP turbine and a double-flow LP turbine.

The main technical data are shown in table 14–4.

Table 14–4 Main technical data of Phu My 3 combined cycle power plant

Plant Configuration:		SCC5-4000F 2×1
Number of blocks		1
Number of gas turbines/block		2
Main fuel		Gas
Ambient Conditions:		
Ambient temperature	°C / °F	30 / 86
Ambient pressure	mbar	1013
Relative humidity	%	82
Cooling type—cooling water temperature	°C / °F	once-through 29 / 84
Performance data per block:		
Total fuel input to gas turbines	MW	1261.2
Gross output	MW	729.4
Gross efficiency (LHV)	%	57.84
Auxiliary consumption and losses	MW	12.60
Net power output	MW	716.8
Net efficiency (LHV)	%	56.84
Heat rate (LHV)	kJ/kWh	6334
Heat rate (LHV)	BTU/kWh	6003
Process heat	MJ/s	0

The 1200-MW Combined-Cycle Power Plant Palos de la Frontera, Spain.

The 3x 400-MW combined-cycle power plant Palos de la Frontera is owned by Union Fenosa Generacion (UFG), a fully owned subsidiary of the Spanish utility Union Fenosa. The plant is located in Andalusia, 100 kilometers southwest of Seville. Handover of the third unit took place in June, 2005.

The Palos de la Frontera CCPP consists of three separately housed Siemens SCC5-4000F combined-cycle blocks, each made up of a gas turbine, generator, and steam turbine in a single-shaft configuration (figure 14–9).

Figure 14–9 View of the Palos de la Frontera combined-cycle power plant

The modularized Reference Power Plant concept developed by Siemens is implemented for the three units. This concept is characterized by optimum standardization of plant systems and components in line with the specific needs of the power plant operators. Special customer requests can be met by using a modularized building-block system.

Costs can thus be saved and the plant can be tailored to the respective customer. A special feature of the single-shaft technology used in the three Spanish plants is the arrangement of the key components. Only one generator is required, which is arranged on a common shaft between the gas and the steam turbine. The single-shaft design allows low investment costs, short construction periods, and highest efficiency levels as well as high operational flexibility (figure 14–10).

Figure 14–10 Layout of the Palos de la Frontera combined-cycle power plant

The gas turbines are Siemens SGT5-4000F (formerly V94.3A) machines. At site conditions, each gas turbine has a power output of 266 MW and an electrical efficiency of 38.6%. The rotor has a hollow shaft with disks that are interlocked via Hirth serrations and axially fixed via a central tie-bolt. The compressor has 15 stages with variable inlet guide vanes for optimized flow control. The turbines feature an annular combustion chamber with 24 low NO$_x$ hybrid burners.

The heat-recovery steam generators (HRSGs) are located outdoors next to the turbine buildings. The HRSGs supplied by Doosan are horizontal, natural circulation units with three pressure levels and reheat. The pipe bundles, which receive the exhaust gas from the gas turbine, are arranged vertically (figure 14–11).

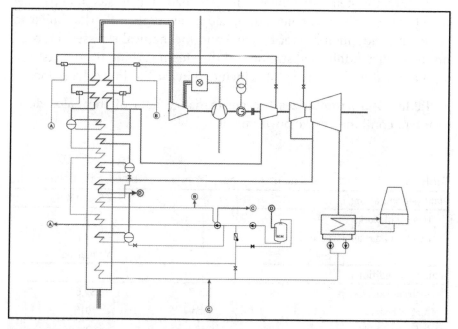

Figure 14–11 Process diagram of the Palos de la Frontera combined-cycle power plant (one unit)

The plant is operating with an average efficiency of 56%. For 2006, the plant achieved an availability of more than 98% (even when accounting for scheduled outages) and produced 7,023 GWh of power. Typically, the plant operates at nearly full load in the summer and winter, but it operates on a stop-start basis in the spring and fall according to the national grid demand and electricity market price.

In today's market, the two main areas of plant operator interest are in achieving minimum load while meeting emission limits and in realizing maximum output with minimum startup time. Accordingly, Siemens has been adapting its technology to improve the cycling capability of the entire plant and not only individual components. Siemens has applied this design philosophy to develop what they refer to as advanced fast cycling (FACY) capability that was first fully tested for hot-start operation at Palos in 2005 and has since been sold into a number of power stations.

Fast hot-start capability (after an eight hour shutdown) means that full load can be achieved in less than 40 minutes. During the startup procedure, the steam turbine is run up to load on cold steam in parallel with the gas turbine. Siemens designed a new type of Benson boiler to cope with the much higher thermal and mechanical stresses imposed on the boiler. Additional stresses on the steam turbine from a fast hot start are within the range that is included in the lifetime calculation.

Table 14–5 summarizes the performance data of the Palos de la Frontera combined cycle power plant

Table 14–5 Main technical data of Palos de la Frontera combined cycle power plant

Plant Configuration:		SCC5-4000F 1S
Number of blocks		3
Number of gas turbines/block		1
Main fuel		Natural gas
Ambient Conditions:		
Ambient temperature	°C / °F	18.3 / 64.9
Ambient pressure	mbar	1015
Relative humidity	%	65
Cooling type		Cooling tower
Performance data per block:		
Total fuel input to gas turbines	MW	673
Gross output	MW	386.6
Gross efficiency (LHV)	%	57.5
Auxiliary consumption and losses	MW	4.7
Net power output	MW	381.9
Net efficiency (LHV)	%	56.8
Heat rate (LHV)	kJ/kWh	6343
Heat rate (LHV)	BTU/kWh	6012
Process heat	MJ/s	0

The Arcos III Combined-Cycle Plant, Spain

Arcos de la Frontera Grupo III in Cadiz, Spain, is the third in an assembly of plants that make up a significant investment and largest power plant in the 100-year history of Spanish energy producer, Iberdrola Generacion, S.A.—also owner and operator of the plant.

In commercial operation since March 2006, Arcos III generates more than 820 MWnet of power for the Spanish electricity grid. The plant is located just between the city of Sevilla and the Port of Algeciras, a heavy industrial area with a high demand for energy.

Arcos I and II began commercial service in January 2005, and each of the groups' 109FA (single-shaft units), together with Arcos III, has a common infrastructure with an administration building/control room, 400kV substation, demineralized and makeup water systems, effluents plant, compressed air and auxiliary steam, natural gas regulation station, and other components. The gas turbine, steam turbine, and generator of each combined-cycle unit are located inside a building where the condenser is also housed (figure 14–12).

Figure 14–12 View of the Arcos combined-cycle power plant

The GE 209FB (2-to-1 multishaft) combined-cycle configuration at Arcos III consists of two GE frame 9FB gas turbine generators, one GE 209D steam turbine, three GE 330H generators, and two heat recovery steam generators. The steam turbine has a single high-pressure/intermediate-pressure (HP/IP) section and one dual-flow low-pressure (LP) section with control and shutoff valves for each sections. The HP steam supply is at 1800 psia (124 bar) and 1050°F (565°C) and the exhaust pressure is 1.73 inches Hg (0.059 bar).

Using natural gas, Arcos III is operating at a combined-cycle efficiency of around 58%, placing it among the world's most efficient combined-cycle power stations. The plant reached another milestone in October 2007, when it surpassed 8000 hours of operation.

Arcos III was the commercial launch site for GE's 50 hertz, 9FB gas turbine technology, a project commissioned by Iberdrola Ingeneria y Construcción. The frame 9FB is among the world's most advanced, air-cooled, 50-hertz gas turbines. Addressing the need for cleaner power, the frame 9FB gas turbines are equipped with GE's advanced, Dry Low NO_x 2+ combustion systems, which limit NO_x emissions to 25 parts per million or less. Table 14–6 summarizes the performance of Arcos III combined-cycle power plant.

Table 14–6 Main technical data of Arcos III combined-cycle power plant

Plant Configuration:		STAG 209FB
Number of blocks		1
Number of gas turbines		2
Number of steam turbines		1
Main fuel		Natural gas
Ambient Design Conditions:		
Ambient temperature	°C / °F	17.9 / 64.2
Ambient pressure	mbar	1005
Relative humidity	%	69
Cooling type		Cooling tower
Performance data per block:		
Total fuel input to gas turbines	MW	1423
Gross output	MW	838
Gross efficiency (LHV)	%	58.9
Auxiliary consumption and losses	MW	15
Net power output	MW	823
Net efficiency (LHV)	%	57.8
Gross heat rate (LHV)	kJ/kWh	6112.5
Gross heat rate (LHV)	BTU/kWh	5794
NO_x emissions	ppm	12
CO emissions*	ppm	0.6

*corrected at 15% O_2

The Arcos III project is designed to produce one-third of the CO_2 emitted by a conventional coal plant and 10 times lower levels of NO_x, while not emitting any SO_2.

The units also feature GE's SPEEDTRONIC™ Mark VI turbine control systems, which offers complete integrated control, protection, and monitoring for generator and mechanical drive applications of gas and steam turbines.

There are currently 43 GE Frame 9FB gas turbines operating or committed for projects worldwide, including 16 in the Iberian region.

The GE 209D steam turbine at Arcos III features 48-inch, last-stage buckets, the industry's largest steel, full-speed (3000 rpm) last-stage buckets in terms of annulus area. Developed by GE and Toshiba, the new buckets are designed for improved plant efficiency and lower electricity production costs.

Arcos III is maintained by Iberdrola with the support of GE through a contractual service agreement for the main equipment.

The Diemen Combined-Cycle Cogeneration Plant, Netherlands

N.V. Energiproduktiebedrijf UNA owns the Diemen 33 combined-cycle power plant that went into operation in the autumn of 1995. The plant is located in the Netherlands close to the town of Muiden, southeast of Amsterdam, and consists of one combined-cycle block with one ABB GT13E2, one HRSG, and a triple-pressure reheat steam turbine providing electrical power and heating to the southeast part of Amsterdam.

In designing the plant, a high degree of flexibility was required, allowing the plant to operate in purely condensing mode with an electrical output of 249 MW or in a combined-power and district-heating mode with 218 MW electrical output and 180 MJ/s district heating production (figure 14–13).

Figure 14–13 View of Diemen combined-cycle cogeneration plant

A high-efficiency triple-pressure reheat cycle was chosen, in which the HP live-steam (89 bar (1276 psig)/505°C (941°F)/49 kg/s (389,000 lb/h)) is fed to the stand-alone HP steam turbine, where it expands and mixes with the IP steam before being reheated in the HRSG (24.5 bar (340 psig)/505°C (941°F)/61 kg/s (484,000 lb/h)) and fed back to the steam turbine. In the IP steam turbine the steam is expanded to the LP level, where it mixes with the LP steam (4.6 bar (52 psig)/saturated/9.3 kg/s (74,000 lb/h)) before undergoing final expansion (figure 14–14).

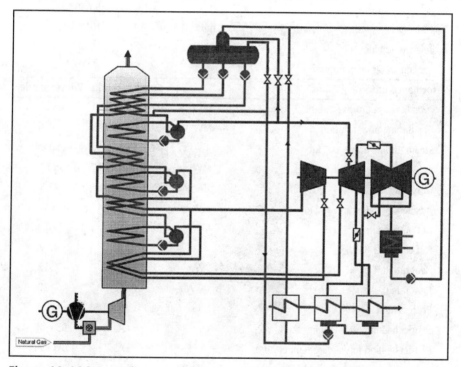

Figure 14–14 Process diagram of Diemen combined-cycle cogeneration plant

To optimize the district-heating mode (winter operation, mainly), three stages of district heating were provided, where the first stage is supplied with a water extraction from the HRSG, and the second and third stages with steam extractions from the steam turbine. For the condensing mode (summer operation, mainly), a double-flow LP steam turbine was chosen to take full advantage of the cold cooling water conditions in the direct cooling system. Table 14–7 summarizes the performance data of the Diemen combined-cycle cogeneration plant.

Table 14–7 Main technical data of Diemen combined cycle cogeneration plant

Plant Configuration:		KA 13E2-1	
Number of blocks		1	
Number of gas turbines/block		1	
Main fuel		Natural gas	
Ambient Conditions:			
Ambient temperature	°C / °F	15 / 59	
Ambient pressure	mbar	1013	
Relative humidity	%	50	
Cooling water temperature	°C / °F	15 / 59	
Performance data:		Summer mode	Winter mode
Total fuel input to gas turbines	MW	455.5	455.5
Gas turbine power output	MW	161.6	161.6
Steam turbine power output		91.1	59.8
Gross output	MW	252.7	221.4
Gross efficiency (LHV)	%	55.5	48.6
Auxiliary consumption and losses	MW	3.5	3.1
Net power output	MW	249.2	218.3
Net efficiency (LHV)	%	54.7	47.9
Heat rate (LHV)	kJ/kWh	6580	7512
Heat rate (LHV)	BTU/kWh	6237	7120
Process heat	MJ/s	0	179.6
District heating return temperature	°C / °F	N.A.	65 / 149
District heating forwarding temperature	°C / °F	N.A.	105 / 211
Fuel utilization	%	N.A.	87.4

The single annular burners of the 13E2 gas turbine restrict NO_x emissions on gaseous fuels to below 25 vppm (15% O_2 dry). To further reduce the NO_x emissions at part load, LP steam is supplied to a heat exchanger at the gas turbine air intake allowing the gas turbine to be operated at nominal TIT for an even wider load range. For temperatures between −7 and +7°C, the heat exchanger is operated to avoid icing at the compressor inlet, which again gives a better efficiency than conventional anti-icing systems where hot air is extracted from the compressor and fed back to the gas turbine air intake.

To give maximum weather protection to the gas turbine, HRSG and steam turbine are arranged indoors. The gas turbine and HRSG are floor mounted and the steam turbine generator unit is mounted on a table with the condenser and district heaters situated beneath it. Plant internal electrical consumers are fed from a common electrical room fed from the auxiliary transformers, connected to the common steam turbine and gas turbine three-winding step-up transformer (figure 14–15).

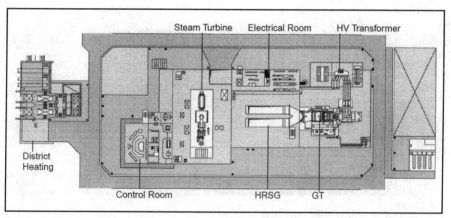

Figure 14–15 General arrangement of Diemen combined-cycle cogeneration plant

The Shuweihat Stage 1 Independent Water and Power Plant (IWPP) in Abu Dhabi, United Arab Emirates

The Shuweihat S1 Independent Water and Power Plant is owned by the Shuweihat Project Company, comprised of the Abu Dhabi Water and Electricity Authority (ADWEA), U.S.-based CMS Energy, and

International Power of the UK. It has an electrical capacity of 1500 MW and a daily production capacity of 100 million imperial gallons (MIGD) of potable water (figure 14–16). The plant is located 250 km west of Abu Dhabi city.

Figure 14–16 View of the Shuweihat S1 plant

As the consortium leader, Siemens was responsible for turnkey erection of the plant and supplied five SGT5-4000F gas turbine generator sets, formerly known as V94.3A, two steam turbine generator sets, and all the ancillary systems. Each gas turbine has a gross rating of 222 MW at reference site conditions (46°C ambient and 42% relative humidity). Siemens also supplied the five dual-pressure heat-recovery steam generators. The steam turbines have a gross rating of 254 MW. The heat-recovery steam generators feature supplementary firing to provide a degree of control in steam production independent of gas turbine firing (figure 14–17).

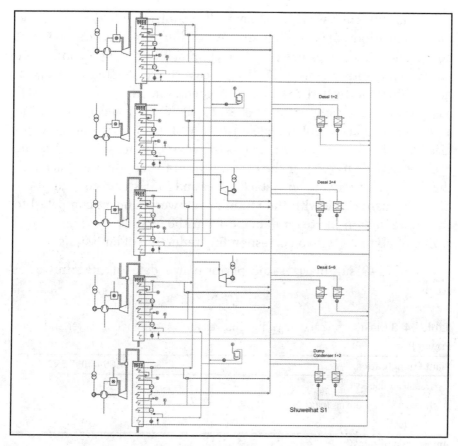

Figure 14–17 Process diagram of the Shuweihat S1 plant

The six multistage flash desalination units (MSF) supplied by Fisia Italimpianti are 30% larger than the previous largest units built. They are supplied with steam from the steam turbine exhaust or from the bypass if the turbines are out of service. The whole desalination facility needs 70 MW of the power capacity to achieve production of 455 million liters of water per day. Ten percent of the seawater feed is converted to potable water with the rest returned to the sea as brine. The resulting distilled water is remineralized to produce potable water that is stored in six storage tanks, which represent 24 hours of water production.

The seven power units have been in commercial operation since August 2004. The whole power plant with all the shared facilities and desalination plant started operation in November 2004.

Special emphasis was given to overall optimization of the power and the desalination parts of the plant. An example of the integration of power and MSF cycles has been the approach to the high condensate return temperature from the MSF process that restricted recovery of heat within the heat-recovery steam generator. In a combined-cycle gas turbine stack temperatures of 90 to 100°C are typical, whereas in a combined cycle supplying steam to a MSF plant it is normally not feasible to achieve temperatures less than 140°C. A modification of the cycle was made to recover the excess heat in the condensate back into the MSF cycle reducing both stack losses and MSF steam consumption. The costs associated with the modification were modest compared to the 6% reduction in steam demand from the MSF process and a fuel saving of up to 2% with a corresponding reduction of emissions.

Table 14–8 summarizes the performance data of the Shuweihat S1 plant.

Table 14–8 Main technical data of the Shuweihat S1 Independent Water and Power Plant (IWPP)

Plant Configuration:		SCC5-4000F 5×2
Number of blocks		1
Number of gas turbines/block		5
Main fuel		Gas
Ambient Conditions:		
Ambient temperature	°C / °F	46/115
Ambient pressure	mbar	1013
Relative humidity	%	42
Cooling type—back pressure	bar/psia	back pressure 2.8/40.6
Performance data per block:		Summer mode Winter mode
Portable water production	MIGD	100
Net power output	MW	1494.9
Net heat rate (LHV)	kJ/kWh	8823
Net efficiency (LHV)	%	40.8
Auxiliary consumption and losses	MW	114.7
Gross power output	MW	1609.6
Gross heat rate (LHV)	kJ/kWh	8194
Gross efficiency (LHV)	%	43.9
Fuel input	MJ/s	3663.8

The 780 MW Repowering Project Vado Ligure, Italy

Pre-existing situation

The Vado Ligure power station is located near Savona, on the seacoast southwest of Genoa; it consists of four thermoelectric conventional units of 320 MW each, commissioned during 1970 and 1971 (see layout in figure 14–18).

Figure 14–18 Vado Ligure (old layout)

The surface of the plant site is of about 42 hectares, including the electric station.

The plant was originally designed to burn residual oil and coal and equipped for oil, but was then equipped to burn coal in the late 1970s. Coal is delivered by ship and carried by enclosed conveyors to the storage pile and to the power plant. Oil is supplied through a pipeline from a dock off-site in an industrial area.

Units 1 and 2 have been unable to burn coal in recent years due to more stringent emission limits: These two units have been considered for combined-cycle repowering. Units 3 and 4, however, have been extensively upgraded to permit continued base load coal burning and meet emission limits regulations.

Each existing unit is independent and made of one boiler, its auxiliaries, one steam turbine and its condenser, independent steam/water cycle and independent connection to the HV station.

The steam turbines are tandem compound reheat units manufactured by Ansaldo, nominally rated at 320 MW.

Condensers are seawater cooled with a once through circulating water system supplied by four 100% capacity vertical single stage pumps.

New Plant Description

Layout

Due to the constraints arising from the existing buildings, a right-angle layout was chosen, as shown in figure 14–19.

One gas turbine and relevant HRSG have been installed in the location of the old unit boiler and ESP unit, previously dismantled by Tirreno Power: their axe is parallel to the one of the steam turbine; the other gas turbine and related HRSG are installed eastward to form a right angle with the first one, in the location of the old workshop/administrative building, partially demolished; the consequence of this solution is that the stacks of the two HRSGs are adjacent one to the other and then structurally tied.

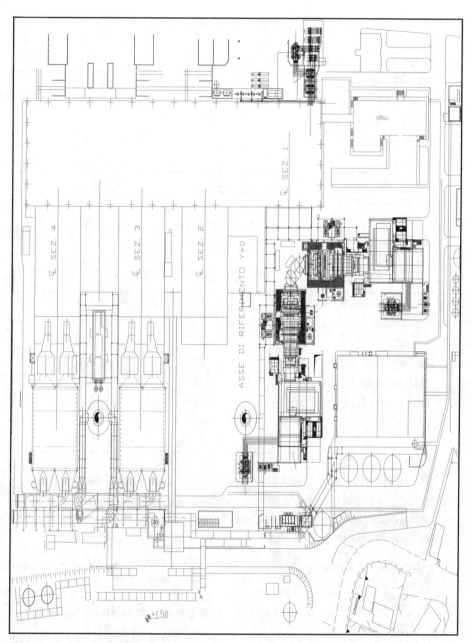

Figure 14–19 Vado Ligure (new layout)

The main technical data are shown in table 14–9.

Table 14–9 Main technical data of Vado Ligure combined cycle power plant

Plant Configuration:		Repowering
Number of blocks		1
Number of gas turbines		2
Number of steam turbines		1
Main fuel		Natural gas
Ambient Design Conditions:		
Ambient temperature	°C / °F	11.6 / 53
Ambient pressure	mbar	981
Relative humidity	%	84
Cooling type		See water
Performance data per block:		
Total fuel input to gas turbines	MW	615
Gross output	MW	359.9
Gross efficiency (LHV)	%	58.5
Auxiliary consumption and losses	MW	6.1
Net power output	MW	353.8
Net efficiency (LHV)	%	57.5
Heat rate (LHV)	kJ/kWh	6261
Heat rate (LHV)	BTU/kWh	5934

Both HRSGs and stacks are installed inside a weather protection cladding, suitably designed for minimizing visual impact. Each gas turbine with relevant generator is installed in buildings that include also their local control system.

The electrical equipments are located in a building adjacent to the turbine hall.

The steam turbine and its electrical generator remain installed in the existing turbine hall.

Extent of the works

Design. In addition to usual design activities, a particular effort was made to optimize layout solutions, check suitability and partial recovery of existing equipment, and avoid interference with the functionality of units 3 and 4, which remained in operation during the conversion.

Supply of main equipment. Main machinery, such as gas turbines, steam turbines, and generators, was manufactured by Ansaldo Energia and transported by ship up to the dock close to the power station.

The two gas turbines are V94.3A2 type, rated 260 MW each; their air intakes are equipped by anti-icing system and inlet air-cooling by fogging. The generators are air cooled.

The steam turbine body is composed by two sections: one IP-HP combined section that replaces the old one and fits the existing pedestals and foundation, and one new inner block of LP installed in the existing LP outer casing; blading is reaction type. Other parts of the old steam turbine were recovered, such as bearing pedestals, cross-over pipe between the LP and IP section, turning gear, lube oil system, steam seal system, and gland air steam condenser. The relevant generator is the old one: A general overhaul was made and the rotor repaired; new auxiliaries were installed, such as excitation system, terminal and neutral point cubicle, current transformers, instrumentation, and so on.

The two HRSGs are horizontal type, with three levels of pressure and reheat. Stacks are 90 meters high and equipped by exhaust gas monitoring system. An architectural cladding was provided for each HRSG and for the coupled stacks.

Steam condenser was subjected to an important overhaul, mainly by replacing water boxes and titanium tube bundles with relevant tube sheets; a new cleaning system has been provided.

Besides the main equipment, the main auxiliary systems, such as feedwater, closed cooling, main steam, circulating water, sampling, chemical injection, and natural gas system are new. The other auxiliary fluids are supplied by the existing power station. Other mechanical equipment were totally or partially re-used after a general overhaul: This is the case, for instance, of condensate extraction pumps, polishing system, vacuum system, sea water piping, and sea water traveling screen. Figure 14–20 shows the thermal process of the repowered plant.

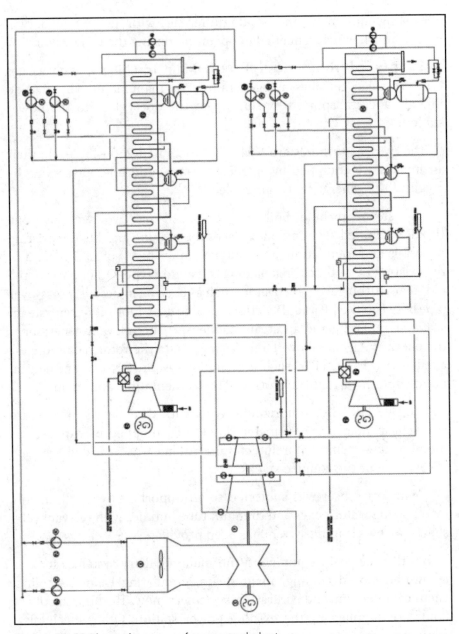

Figure 14–20 Thermal process of repowered plant.

1) Gas turbine
2) Gas turbine generator
3) HRSG
4) Steam turbine
5) Steam turbine generator
6) Condenser
7) Condensate extraction pump
8) Feedwater pump

The supply of the equipment of the electrical system was basically new, and it includes the installation of three 300 MVA, 380KV step-up transformers and a 400kV Gas Insulated Switchgear (GIS) for the connection of the generator units with the TERNA S/S. Figure 14–21 shows a CAD image of the repowered plant.

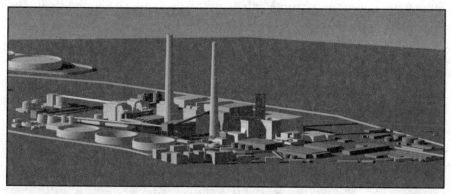

Figure 14–21 CAD Image of the repowered plant

Atmospheric emissions

The new plant has been designed to limit the flue gas emissions to the atmosphere to the following limits:

NO_x	≤ 40 mg/Nm³	
CO	≤ 30 mg/Nm³	Concentration at 15% O_2, vol. dry, at 0°C and 1013 hPa
Particulate	< 1 mg/Nm³	

All emission limits are obtained with the gas turbine operating in dry mode, without injection of steam or water.

Works on site started on June 2005, with old building demolition.

On July 9, 2007, the plant began to produce electricity to face the summer peak. Hand-over to Tirreno Power was done on October 13, 2007.

The Puertollano IGCC plant, Spain

The 298 MW IGCC plant in Puertollano is owned and operated by ELCOGAS, a consortium of eight major European utilities and three technology suppliers. Puertollano is located in the central south part of Spain, 200 kilometers from Madrid, in the province of Ciudad Real. The IGCC is 10 kilometers east-southeast of the town of Puertollano.

The plant started continuous syngas operation in November 1998 and supplies electricity to the Spanish grid utilizing a Siemens gas turbine V94.3 and a Prenflo gasifier. Originally, the Prenflo gasification has been developed by Krupp Koppers in cooperation with Shell, and is, therefore, based on similar design features.

The plant converts about 700,000 tons of mixed fuel per year at full operational capacity. The feedstock is made up of 50% petroleum coke coming from the nearby REPSOL refinery and 50% of high ash coal supplied from the local ENCASUR mine.

The Puertollano IGCC plant is designed as fully integrated concept, which means that the air required for oxygen separation is entirely supplied from the gas turbine compressor (for flow scheme, see figure 14–22).

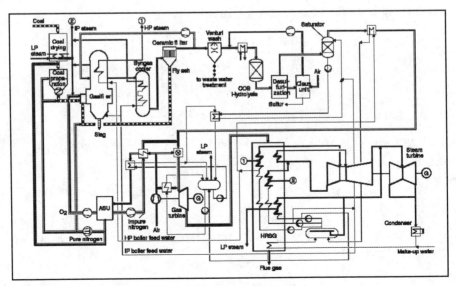

Figure 14–22 Flow scheme of Puertollano IGCC plant

This particular configuration was designed to enable the highest possible efficiency at contract date. It applies standard Siemens V94.2 gas turbine technology equipped with horizontal silo type combustion chambers. Based on local fuel mixture, the plant operates with about 41.5% (LHV) efficiency but considering standard coal quality with lower ash content and fusion temperature an efficiency level of 45% (LHV) could be achieved. Beyond Siemens combined-cycle, the basic technologies are the air separation unit (ASU) delivered from Air Liquide and the gas island consisting of gasifier, fuel preparation, and gas treatment technology provided by Krupp Uhde (former Krupp Koppers). Most of the hardware was delivered from Babcock Wilcox Español.

The ASU supplies the oxygen to the gasification process with a purity of 85 vol% and provides two nitrogen fractions, one for coal feeding and inertisation with a purity of 99.5 vol% and the second waste nitrogen feed for syngas dilution and gas turbine NO_x control.

The syngas generation takes place in an entrained-flow oxygen-blown system with dry fuel feeding, and the hot syngas leaving the reactor is cooled down by HP and IP steam production and then passes several cleaning steps. First particles, halogens, and alkalines are removed, and subsequently COS and H_2S are washed by MDEA scrubber before entering the gas turbine fuel gas conditioning system for preheating and dilution.

The gas turbine is equipped with a syngas diffusion burner derived from a standard hybrid burner and capable to operate with natural gas as secondary fuel. Before entering the combustion chamber, the syngas is saturated with water vapor and diluted with waste nitrogen to keep NO_x emission well below 60 mg/Nm³ (15% O_2).

Although a high-sulfur-containing fuel mixture is processed, the plant emissions are only 10mg/Nm³ of SO_2, which is far below the current limitations for conventional steam power plants of about 160mg/Nm³ SO_2 with 15% O_2.

On the water steam side, a high efficient triple-pressure reheat cycle was chosen with HP life steam at 127 bar, reheat life steam at 35 bar, and LP steam at 6.5 bar.

An overview of the main technical data of the IGCC Puertollano is given in figure 14–23.

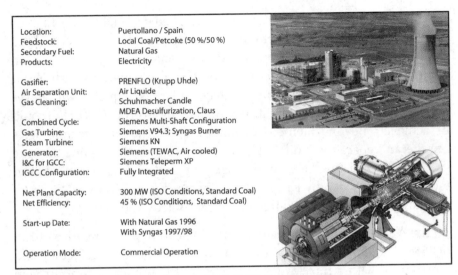

Figure 14–23 Main technical data of the IGCC Puertollano

Because of the first of its kind technology character the Puertollano plant struggled with a number of availability issues at the beginning of operation. These initial design issues have been solved now. A major conclusion, however, was that availability (see figure 14–24) suffers due to the fully integrated concept chosen and the higher complexity of the plant, which resulted from it.

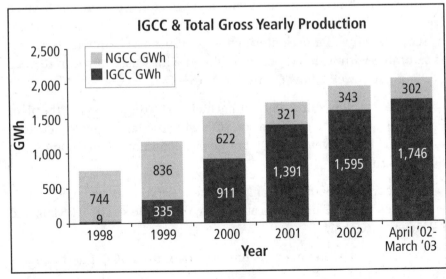

Figure 14–24 Availability between 1996 and 2003

The Monthel Cogeneration Plant, Switzerland

The 121-MW (66 MW thermal plus 55 MW electrical power at 100%-GT and 100%-HRSG supplementary firing) Monthel cogeneration power plant is owned by Monthel SA, which is owned entirely by Alpiq group, one of the leading Swiss electrical utility companies, and supplies process steam and part of the produced electricity to the chemical companies on site and is connected to the local medium voltage grid to supply the rest of the electricity to Swiss electricity consumers.

The plant is located about 20 km south of Montreux on the chemical site of Monthey 400 meters above sea level. The plant will be fired with natural gas only taken from the nearby high pressure line.

The power plant consists of a two-shaft block with one Siemens SGT-800 gas turbine, one Aalborg HRSG with supplementary firing, and one Siemens steam turbine with process steam extraction, as well the necessary auxiliary equipment. The steam, which is not used as process steam, is condensed in a river water-cooled condenser.

An efficient double-pressure cycle was chosen for the steam cycle. The high pressure live steam data are 64 bar/525°C and the low pressure data are 8.5 bar/210°C.

Process steam is delivered to the chemical plant at three different pressure levels:

- 56 bar
- 12 bar
- 6 bar

The process diagram of the plant is shown on figure 14–25.

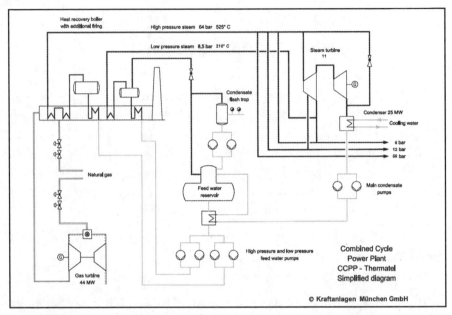

Figure 14–25 Process diagram of the Monthel cogeneration power plant

The Monthel plant has the gas turbine with its generator and the HRSG as an outdoor arrangement. The steam turbine with its generator and the auxiliary equipment is located in a machine house adjacent to the gas turbine/HRSG train (figure 14–26).

Figure 14–26 Plant arrangement (Kraftanlagen Muenchen GmbH)

The condenser is below the steam turbine, which is placed on a table.

The arrangement is very compact, with a footprint of the plant measuring about 50 by 80 meters.

This plant will be the first of this kind and size in Switzerland. Due to special local emission regulations an SCR will be installed in the HRSG to further reduce the NO_x emission level of the gas turbine down to approximately 23 mg/Nm².

The plant is built under a fixed lump sum turnkey contract by the German plant engineering company KAM (Kraftanlagen Muenchen GmbH).

The energy services company CIMO (Compagnie Industrielle de Monthey SA), which supplies steam, electricity, and other services to the chemical companies on the Monthey site, will be responsible for the operation and maintenance of the plant in the frame of the Operation & Maintenance contract which will be signed between the owner Monthel SA and CIMO SA.

Table 14–10 summarizes the performance data of the Monthel cogeneration plant

Table 14–10 Main technical data of Monthel cogeneration plant

Plant Configuration:		
Ambient Design Conditions:		
Ambient temperature	°C / °F	10/50
Ambient pressure	mbar	966
Performance data at 100 %-GT and 100 %-HRSG supplementary firing:		
Net electrical output	MW	54.6
Gas turbine output (gross)	MW	44.2
Steam turbine output (gross)	MW	11.0
Net thermal output	MW	65.6
Auxiliary power	MW	0.6
Fuel heat input (LHV)	MW	141.5
Max. net electrical output at min. steam load at 100 %-GT and 100 %-HRSG supplementary firing	MW	57.4

15 Conclusion

The thermodynamic advantages of the combined-cycle concept enable efficiencies to be reached far above those of other types of thermal power plants. This technology above all others can fully exploit the high-temperature potential of modern gas turbines and the low-temperature cold end of the steam cycle. Coupled with low investment and operating costs, short delivery times, and high operational flexibility, these factors ensure an overall low cost of electricity.

The thermodynamic advantages are also beneficial in cogeneration applications, especially where high electrical output is required because of the stable electrical output contribution of the gas turbine, which is not influenced by the steam process. The possibility of supplementary firing in the HRSG provides even greater operational flexibility where variations in the steam extraction demand are required.

Combined-cycle plants are suitable for daily cycling operation due to short startup times as well as for continuous baseload operation. Part-load efficiencies are also high due to the control of the gas turbine mass flow using variable inlet guide vanes in the compressor.

Combined cycles can be cooled by a cooling tower, a direct water-cooling system, or air-cooled condensers ensuring a wide range of applications. Where water is scarce, they are advantageous because the cooling requirement per unit of electricity produced is low because the main cooling requirement applies only to the steam process (one-third of total output).

Combined-cycle technology is one of the most environmentally acceptable large-scale power plant technologies in use today. Fuelled mainly by natural gas, they are ideally suited for use in heavily populated regions due to low emission levels for pollutants such as nitrous oxides (NO_x), carbon monoxide (CO), and particles. High efficiency means less fuel must be burned for each unit of electricity produced, which also contributes to these low emission levels.

The carbon dioxide emissions (CO_2) are the lowest of all fossil-fired power plants. In a modern gas-fired combined-cycle plant they are only 40% of those of a coal-fired plant with the same output. This is, of course, a great advantage with today's awareness of climatic change as well as an economical advantage when CO_2-emission certificates have to be acquired.

Today's combined-cycle plants have reached a very high technical standard: net efficiencies of up to 59% are achievable. Block sizes of over 400 MW are state of the art. Nevertheless the development in combined-cycle plants continues through the advancement of the constituent components, which are now being developed specifically for use in combined-cycle applications. This has influenced the gas turbine in particular, with a trend towards pressure ratios and turbine inlet temperatures optimized for combined-cycle applications and a move towards more innovative machine concepts such as sequential combustion or even steam cooling.

Although the fuel flexibility of combined-cycle plants is limited to gases and some oils, this is becoming less significant as global gas distribution increases bringing natural gas to many countries that do not have their own natural reserves. Gas is either transported through pipe or shipped as liquid (LNG).

A wide range of combined-cycle power plant concepts is available and selection of a cycle concept is made according to the criteria of a specific project. Now that the combined-cycle is established as one of the main global power generation technologies, ideal cycle solutions are emerging matched to certain sets of criteria. Manufacturers have recognized in this a potential for the development of a range of standard plants. This leads to shorter delivery times, faster permitting, and lower risk to the investor due to proven components and systems.

These and other factors point to a continuing dependence on combined-cycle technology for generating a main part of the world's electrical energy well into the future. Especially if the concept of carbon capture and storage (CCS) gets imposed, combined-cycle plants with integrated coal gasification (IGCC) could be the standard coal fired stations of the future.

Appendix A

Calculation of the Operating Performance of Combined-Cycle Installations

When determining the performance data of a combined cycle, the gas turbine data are usually given and only the HRSG and steam turbine can be individually calculated. This appendix shows the main steps involved in such a calculation process, starting with the HRSG.

1. Equations for the heat exchangers of the HRSG

The equations of energy, impulse, and continuity are used to calculate the steady-state behavior of economizers.

The continuity equation comes down in the steady state to:

$$\sum \dot{m} = 0 \tag{A-1}$$

The impulse equation can be simplified into:

$$\Delta p = f(\text{geometry}) \tag{A-2}$$

However, because the pressure loss both in the economizer and in the evaporator has a negligible influence on the energy equations, the assumption

$$\Delta p = 0 \tag{A-3}$$

is valid. In this case, the pressures along the heat exchanger remain constant on both the gas and water sides. The energy equation for a small section dx of a heat exchanger, which can be treated approximately as a tube, can be written as follows:

$$d\dot{Q} = k \cdot \Delta t \cdot \pi \cdot d \cdot dx \tag{A-4}$$

If it is assumed that the heat transfer coefficient k remains constant over the entire length of the heat exchanger (economizer or evaporator), equation A–4 becomes:

$$\dot{Q} = k \cdot S \int_0^L \Delta t(x) \tag{A-5}$$

In general, the expression cannot be integrated. The heat exchanger must, therefore, be dealt with in small elements.

In the special cases of a heat exchanger with counter or parallel flow, however, integration is possible assuming that the specific heat capacities of both media along the heat exchanger remain constant.

The result of the integration is the logarithmic average value for the difference in temperature, which can be written in the form:

$$\int_0^L \Delta t(x) = \frac{\Delta t_{Inlet} - \Delta t_{Outlet}}{\ln\left(\dfrac{\Delta t_{Inlet}}{\Delta t_{Outlet}}\right)} = \Delta t_m \tag{A-6}$$

This average value can also be used for a superheater, an evaporator, or a recuperator. The heat exchangers do not, in fact, operate in accordance with an ideal counterflow principle, but the errors remain negligible.

Substituting equation A–6 into equation A–5 yields:

$$\dot{Q} = k \cdot S \cdot \Delta t_m \tag{A-7}$$

From equation A–1, the amount of heat exchanged can be expressed as follows:

$$\dot{Q} = \dot{m}_S \cdot \Delta h_S = \dot{m}_G \cdot \Delta h_G \tag{A-8}$$

At the design point, equations A–7 and A–8 become:

$$\dot{Q}_0 = k_0 \cdot S \cdot \Delta t_{m0} \tag{A-9}$$

$$\dot{Q}_0 = \dot{m}_{S0} \cdot \Delta h_{S0} = \dot{m}_{G0} \cdot \Delta h_{G0} \tag{A-10}$$

Dividing equation A–7 by equation A–9 and equation A–8 by A–10 yields the formulae:

$$\frac{\dot{Q}}{\dot{Q}_0} = \frac{k \cdot \Delta t_m}{k_0 \cdot \Delta t_{m0}} \tag{A-11}$$

$$\frac{\dot{Q}}{\dot{Q}_0} = \frac{\dot{m}_G \cdot \Delta h_G}{\dot{m}_{G0} \cdot \Delta h_{G0}} \tag{A-12}$$

Subtracting equation A–12 from A–11 produces:

$$\frac{\Delta t_m}{\Delta t_{m0}} = \frac{k_0 \cdot \dot{m}_G \cdot \Delta h_G}{k \cdot \dot{m}_{G0} \cdot \Delta h_{G0}} \tag{A-13}$$

This is the nondimensional, global equation of heat transfer for the heat exchanger. If, in addition, equation A–8 is taken into consideration and the heat transfer coefficient k is known, a system of equations is obtained that defines the heat exchanger.

2. Finding the heat transfer coefficient of HRSG sections

The heat transfer coefficient can be calculated using the following equation:

$$k = \frac{1}{\dfrac{1}{\alpha_G} + \dfrac{d_1}{2 \cdot \lambda} \cdot \ln\left(\dfrac{d_1}{d_2}\right) + \dfrac{d_1}{d_2 \cdot \alpha_S}} \tag{A-14}$$

However, the relative values k/k_0 appear in the heat transfer equation. From this:

$$K = \frac{k}{k_0} = \frac{\dfrac{1}{\alpha_{G0}} + \dfrac{d_1}{d_2 \cdot \alpha_{S0}}}{\dfrac{1}{\alpha_G} + \dfrac{d_1}{d_2 \cdot \alpha_S}} \tag{A–15}$$

The heat transfer coefficients on the gas side of the economizer and the evaporator (α_G) are from 0.1 to 0.01 times as large as those on the steam side (α_S). Moreover, both values always shift in the same direction (++, --).

For these reasons, the following approximation can be used:

$$K = \frac{k}{k_0} = \frac{\alpha_G}{\alpha_{G0}} \tag{A–16}$$

The α-value on the gas end can be calculated as follows using the Nusselt number:

$$Nu_G = c \cdot Re^m \cdot Pr^n = \frac{\alpha_G \cdot d_1}{\lambda_G} \tag{A–17}$$

Here, c, m, and n are constants that depend mainly upon the geometry involved. From this, the following expression is obtained:

$$\alpha_G = c' \cdot \lambda_G \cdot Re^m \cdot Pr^n \tag{A–18}$$

If this is substituted into equation (A–16), the geometric constant c' disappears:

$$K = \frac{\lambda \cdot Re^m \cdot Pr^n}{\lambda_{G0} \cdot Re_0^m \cdot Pr_0^n} \tag{A–19}$$

For gases, the Prandtl number is almost exactly a constant; therefore:

$$K = \frac{\lambda}{\lambda_{G0}} \cdot \left(\frac{Re}{Re_0}\right)^m \tag{A–20}$$

For the Reynolds number, the following expression applies:

$$\text{Re} = \frac{c_G \cdot \rho_G \cdot d_1}{\mu_G} \qquad (A-21)$$

By substituting $\dot{m}_G/S$ for $c_G \cdot \rho_G$, one obtains:

$$\text{Re} = \frac{\dot{m}_G \cdot d_1}{\mu_G \cdot S} \qquad (A-22)$$

Then, substituting this expression into equation (A–20), the geometric parameters disappear:

$$K = \frac{\lambda_G}{\lambda_{G0}} \cdot \left(\frac{\dot{m}_G \cdot \mu_{G0}}{\dot{m}_{G0} \cdot \mu_G}\right)^m \qquad (A-23)$$

If the mass flow is constant, all that remains is:

$$K = \frac{\lambda_G}{\lambda_{G0}} \cdot \left(\frac{\mu_{G0}}{\mu_G}\right)^m \qquad (A-24)$$

For m, one can use 0.57 for HRSG tubes that are staggered and 0.62 for inline pipes.

The value of the expression $K = \frac{\lambda_G}{\lambda_{G0}} \cdot \left(\frac{\mu_{G0}}{\mu_G}\right)^m$ does not vary notably and depends mainly on the properties of the gas. It can be replaced with the following approximation:

$$\frac{\lambda_G}{\lambda_{G0}} \cdot \left(\frac{\mu_{G0}}{\mu_G}\right)^m = 1 - (\bar{t}_0 - \bar{t}) \cdot 5 \cdot 10^{-4} \text{ (in SI-Units)} \qquad (A-25)$$

$\bar{t}_0$ and $\bar{t}$ are the average gas temperatures along the heat exchanger in the design and operating point. This produces the relative value of K:

$$K = \left(\frac{\dot{m}_G}{\dot{m}_{G0}}\right)^m \cdot 1 - (\bar{t}_0 - \bar{t}) \cdot 5 \cdot 10^{-4} \qquad (A-26)$$

In this equation, only the exponent m depends to a slight extent on the boiler geometry.

It is more complicated to calculate an exact K-value for the superheater because the heat transfer on the steam side is lower than that in the evaporator.

As soon as all these equations are available for all parts of the boiler, the HRSG is defined mathematically. Similar equations can also be formulated for calculating the condenser. With this methodology it is possible to calculate off-design points of operation without having to know the actual geometry of the boiler; only the design point has to be known.

3. The steam turbine

Most steam turbines in combined-cycle plants operate in sliding pressure mode and have no control stage with nozzle groups. This simplifies calculations because simulation of the control stage and the inlet valves is fairly complicated.

A portion of a steam turbine with no extraction is defined by one equation for its swallowing capacity and one for its efficiency. The swallowing capacity can be defined using the Law of Cones (ellipse law).

$$\left(\frac{\dot{m}_S}{\dot{m}_{S0}}\right) = \frac{\overline{v} \cdot p_\alpha}{v_0 \cdot p_{\alpha 0}} \sqrt{\frac{p_{\alpha 0} \cdot v_{\alpha 0}}{p_\alpha \cdot v_\alpha}} \cdot \sqrt{\frac{1 - \left(\frac{p_\omega}{p_\alpha}\right)^{\frac{n+1}{n}}}{1 - \left(\frac{p_{\omega 0}}{p_{\alpha 0}}\right)^{\frac{n+1}{n}}}} \qquad \text{(A–27)}$$

In condensing steam turbines, the pressure ratio is always very small due to the low pressure at the steam turbine exhaust p_ω. If simplified, this makes it possible to replace the quadratic expression with 1. The ratio of the swallowing capacities is as well close to 1.

What remains is:

$$\left(\frac{\dot{m}_S}{\dot{m}_{S0}}\right) = \sqrt{\frac{p_\alpha \cdot v_{\alpha 0}}{p_{\alpha 0} \cdot v_\alpha}} \qquad \text{(A–28)}$$

At a constant rotational speed, the efficiency of a stage depends only upon the enthalpy drop involved. In part-load operation, however, no important change occurs in that drop except in the last stages.

Because this means that the greatest portion of the machine is operating at a constant efficiency, it can be assumed that the polytropic efficiency remains constant. The turbine efficiency is calculated in the same way as the design point.

The following formulae are used to calculate efficiency. For parts of the turbine operating in the superheated zone:

$\eta_{pol,dry} = $ constant

For parts in the wet steam region:

$$\eta_{pol} = \eta_{pol\text{-}dry} - \frac{(1 - x_\alpha) + (1 - x_\omega)}{2} \tag{A-29}$$

The polytropic efficiency selected should be such that the design power output is once again actually attained in the design point.

The following equation is used to determine the isentropic efficiency:

$$\eta_{is} = \frac{1 - \left(\frac{P_\omega}{P_\alpha}\right)^{\frac{x-1}{x}\eta_{pol}}}{1 - \left(\frac{P_\omega}{P_\alpha}\right)^{\frac{x-1}{x}}} \tag{A-30}$$

These equations make it possible to establish the expansion line of the steam turbine. The power output of the steam turbine can be determined from this considering dummy piston, steam turbine exhaust, mechanical, and generator losses. The actual losses are based on the steam turbine type, size, and live-steam pressure. As a guidance the following losses can be considered:

- Piston losses
  For reaction type steam turbines, the piston losses account for 400 to 1000 kW mechanical equivalent losses. On the other hand impulse type steam turbines by their nature do not need a piston to compensate the axial thrust and would only have gland steam losses accounting for approx. 0.2% of the mechanical output.

- Steam turbine exhaust losses
  Steam velocity at the steam turbine exhaust cause enthalpy losses, generally in the range of 20 to 35 kJ/kg

- Mechanical losses
  Bearing losses and other mechanical losses normally account for 0.3% mechanical losses

- Generator losses
  The generator efficiency varies from 97 to 99% dependent on the steam turbine size

4. Solving the system of equations

All the equations in the HRSG, the steam turbine, and so on create a system that can be solved only by iteration.

The following values are known:

- Thermodynamic data at the design point
- The marginal conditions for the particular operation to be calculated (exhaust data for the gas turbine, cooling water data, etc.)
- Operating mode of the feedwater tank (sliding or fixed pressure)
- Gas and steam tables

The following information must be found:

- Behavior of the steam cycle

Figure A–1 shows the method to find the solution. Starting with the superheater a first estimate for live-steam temperature and pressure is made. Using the Law of Cones and the energy equation, the live-steam flow and the gas temperature after the superheater are found. Next, from the heat transfer equation, a new value for live-steam temperature can be determined. This is then used for further iteration. The procedure is repeated until all three equations have been solved.

The energy and heat transfer equations for the economizer and the evaporator can be used to determine a second approximation for live-steam pressure.

If the feedwater tank is in sliding pressure operation, a first estimate for feedwater temperature is also necessary.

The new value obtained for live-steam pressure is then used to continue calculation of the superheater and the turbine until all equations for the boiler and the Law of Cones agree. The next step is to calculate the preheating of the feedwater. This is used to find a new approximation for feedwater temperature if the pressure in the feedwater tank varies. The boiler is then recalculated with this new value. Finally, the condenser pressure and extraction flow are determined in another iteration. Then, from these new values, the power output of the steam turbine can be determined.

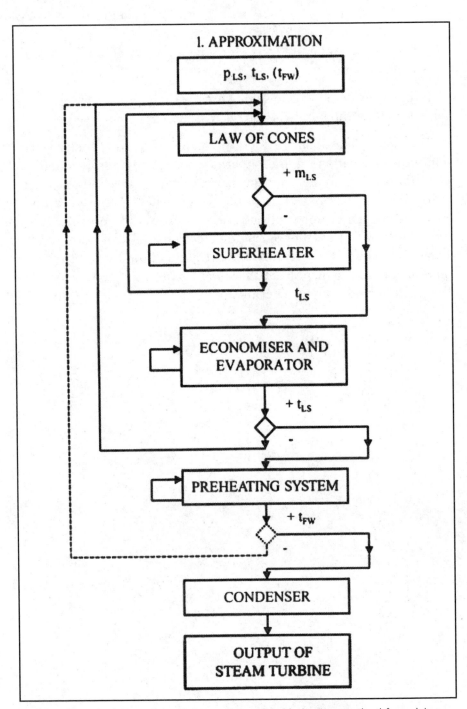

Figure A–1 Calculation of operating and part-load behavior: method for solving the system of equations

Appendix B

Conversions Table

Conversion of the main units used in this book.

Multiply	by	to obtain
bar	14.5	psia (psig=psia-14.5)
Btu	1.055	kJ
ft	0.30480	m
gal (U.S.)	3.7854	l
inch	2.54	cm
kJ	0.94781	Btu
kg	2.20046	lb
l	0.26417	U.S. gal
m	3.2808	ft
psia	0.069	bar

Conversion formulae

To convert	into	formula
°C	°F	$(9/5)°C + 32$
°F	°C	$5/9 (°F - 32)$
K	°C	$K - 273.15$
K	°R	$1.8K$

Appendix C

Symbols Used

- c   Velocity
- d   Diameter of a tube or pipe
- Δh Enthalpy difference
- k   Heat transfer coefficient
- $\dot{m}$ Mass flow
- Nu   Nusselt number
- P   Pressure
- Δp Difference in pressure
- Pr   Prandtl number
- $\dot{Q}$ Heat flow, amount of heat
- Re   Reynolds number
- S   Surface area
- t   Temperature in °C
- Δt Difference in temperature
- υ Specific volume
- x   Steam content of the wet steam

α Heat transfer coefficient

η Efficiency

μ Dynamic viscosity

λ Heat conductance

$\bar{\nu}$ Average swallowing capacity of the turbine

ρ Density

Indices Used

A   Air

FW   Feedwater

G   Flue gas

IS   Isentropic

LS   Live steam

Pol   Polytropic

S   Steam

0   Design point

1   Inlet/inside

2   Outlet/outside

α Section of a turbine inlet

ω Section of a turbine outlet

16 Bibliography

Chapter 4

Knizia, K.: "Die Thermodynamik des Dampfprozesses," 3rd ed., vol. I, Springer Verlag.

Chapter 5

Bachmann, R., M. Fetescu, and H. Nielsen. "More than 60% Efficiency by Combining Advanced Gas Turbines and Conventional Power Plants." PowerGen 95, Americas.

Balling, L., E. Wolt, and G. Baumgärtel. "Modular Power Plant Design, the key to success for economic power plant engineering." Power-Gen Europe, Brussels, 2001.

Jury, Dr. W. "Single-Shaft Power Trains (SSPTs): Flexible and Economic Power Generation for the Future." PowerGen Asia, Hong Kong (September 1994).

Kail, C., B. Rukes. "Fortschrittliche Gas- und Dampfturbinenprozesse zur Wirkungsgrad- und Leistungssteigerung bei GUD-Kraftwerken." VDI Tagungsband 1182/1995.

Rukes, B., R. Taud. "Perspectives of Fossil Power Technology." VGB PowerTech 10/2002: 71–76.

Rukes, B., R. Taud. "Status and Perspectives of Fossil Power Generation." ENERGY 29/2004.

Rukes, B., L.Balling. "Entwicklung, Bau, Service und Betrieb von Kraftwerken." BWK-Das Energie-Fachmagazin-, Bd. 59/2007, Nr.1/2: 69–74.

Warner, J., H. Nielsen. "A selection method for optimum combined-cycle design." ABB Review, 8/93.

Chapter 6

Baerfuss, P.A., K.H. Vonau. "GT8C single shaft combined-cycle: First Application of Industrial Power company Cogeneration at Eastern Industrial Estate, Thailand." Power-gen '96 ASIA, New Dehli, India (September 1996).

Oest, H. "Comparison between the bombined-cycle and the HAT Cycle" thesis. Department of Heat and Power Engineering, T Lund Institute of Technology, (August 1993).

Pfost, Rukes, B. "Gas Turbines Increase Power and Efficiency of Steam Power Plants." Power-Gen Europe, 1998.

Chapter 7

Kiesow Dr. H.J., D. Mukherjee. "The GT24/26 Family Gas Turbine: Design for Manufacturing." British Engineer's Conference, U. K. (July 1997).

Lenk, U., P. Voigtländer. "Use of Different Fuels in Gas Turbines." VGB PowerTech 8/2001.

Leusden, Christoph Pels, Christoph Sorgenfrey, and Lutz Dümmel. "Performance Benefits Using Siemens Advanced Compressor Cleaning System." ASME Paper 2003-GT-38184.

Nag, Pratyush, Matthew LaGrow, Jianfan Wu, Khalil Abou-Jaoude, and Jacqueline Engel. "LNG Fuel Flexibility in Siemens' Land-Based Gas Turbine Operations." Electric Power Conference – Chicago, IL, May 1–3, 2007.

"GT26 kicks off New Zealand combined-cycle program." Reprint from Turbomachinery International, March-April 1997.

Chapter 8

Bender, Grühn, Scheidel Rukes. "Online-Ferndiagnose von Turbosätzen und GUD-Gesamtanlagen." VDI Tagungsband 1641/2001.

Chapter 9

Frutschi, H.U., A.Plancherel. "Comparison of Combined-cycles with Steam Injection and Evaporisation Cycles." 1988 ASME Cogen-Turbo IGTI, vol. 3.

Kehlhofer, R. "Calculation of Part-Load Operation of Combined Gas/Steam Turbine Plants." Brown Boveri Review 65, 1978 (10), p. 679.

"Performance test methods to determine the benefits of various GT modernizations & upgrades." 2006 ASME Power – PWR 2006-88056.

"Performance test code on overall plant performance." ASME PTC 46 1996.

"Performance test code on gas turbines." ASME PTC 22-2005:

"Performance test code for HRSG." ASME PTC 4.4-1981.

"Performance test code for steam turbines in combined cycles." ASME PTC 6.2-2004.

"Acceptance test code for combined cycle power plants." ISO 2314 – Amendment 1–1997.

Chapter 11

Emberger, H., D. Hofmann, and C. Kolk. "Economic Evaluation of Cycling Plants- An Approach to Show the Value of Operational Flexibility." Power-Gen Europe, 2006.

Henkel, N., E.Schmid, and E. Gobrecht. "Operational Flexibility Enhancements of Combined Cycle Power Plants." Power-Gen Europe, June 26–28, 2007.

Hofmann, D., G. Winkler. "SCC5-4000F 2x1 – A Comeback of the Multi-Shaft Configuration to the Market." Power-Gen Asia, 2007.

Kail, C., B. Rukes, W. Märker, F. Strobelt, and I. Weber. "Leistungssteigernde Maßnahmen bei GUD-Kraftwerken." BWK-Das Energie-Fachmagazin-, Bd. 59/2007, Nr. 12: 47-52.

Riedle, K., W. Drenckhahn, P. Kluesener, and I. Pyc. "Clean Fossil Technologies-Attractive Options for Creating Customer Value in Future Power Plants." World Energy Conference, September 2004, Sydney.

Chapter 12

Ratafia-Brown, J., L. Manfredo, J. Hoffmann, M. Ramezan, "Major Environmental Aspects of Gasification-Based Power Generation." U.S. Department of Energy, 2002.

Higman, C., M. van der Burgt. "Gasification." Elsevier Science, 2003.

Wolf, K.J., W. Renzenbrink, J. Ewers "RWE Clean Coal Programme – IGCC Power Plant with CO_2 Capture & Storage." 8th European Gasification Conference, 2007.

BINE Informationsdienst projektinfo, Kraftwerke mit Kohlevergasung, 09/2006.

Hannemann, F., U. Schiffers, J. Karg, M. Kanaar. V94.2 Buggenum Experience and Improved Concepts for Syngas Application, GTC 10/2002.

Gadde, S., J. Wu, A. Gulati, Mc Quiggan, B. Koestlin, B. Prade. "Syngas Capable Combustion Systems Development for Advanced Gas Turbines." 2006.

Fossil Energy Power Plant Desk Reference, DOE/NETL-2007/1282.

SFA Pacific, Inc., Gasification. Critical Analysis of Technology, Economics, Markets, 2007.

Whysall, M. "Gasification in Search of Efficiency." The 8th European Gasification Conference September, 2007.

CASTOR-ENCAP-CACHET-DYNAMIS Common Technical Workshop January 2008-Lyon, FRANCE.

IGCC Combined-cycle Power Plants: ABB Power Generation publication PGT 97 E.

Kehlhofer, R., B.R. Thompson, C. Greil. "Coal Gasification Combined-cycle Plants: A Clean Way from Coal to Electricity." VGB Congress, Strassbourg, France, 1986.

Chapter 13

IPCC Special Report on Carbon Dioxide Capture and Storage, 2005. Prepared by Working Group III of the Intergovernmental Panel of Climate Change. IPCC, 2005.

Climate Change 2007: Mitigation of Climate Change. Working Group III contribution to the Intergovernmental Panel on Climate Change Fourth Assessment Report. IPCC, 2007.

Feron, P. H. M., et al. Development of post-combustion capture of CO_2 within the CASTOR Integrated Project: Results from the pilot plant operation using MEA. 3rd Int. Conf. on Clean Coal Technologies for our Future, 15–17 May 2007, Cagliari/Sardinia.

Göttlicher, G. "Capture, transport and storage of CO_2 from fossil-fired power plants." VGB PowerTech 5/2003, pp 96–101.

Hustad, C. W., M. Austell. "Mechanisms and incentives to promote the use and storage of CO_2 in the North Sea." European Energy Law Report, 2004.

Kvamsdal, H. M., K. Jordal, and O. Bolland. "A quantitative comparison of gas turbine cycles with CO_2 capture." Energy 32 (2007), pp 10–24.

Viebahn, P., et al. "Comparison of carbon capture and storage with renewable energy technologies regarding structural, economic, and ecological aspects in Germany." Int. J. Greenhous Gas Control, 2007, doi: 10.1016/S1750-5836(07)00024–2.

World Energy Council. The World Energy Book. Issue 3, 2006.

The European Technology Platform for Zero Emission Fossil Fuel Power Plants (ZEP). Strategic Research Agenda, 2006.

Chapter 14

Hafsmo, J., T. Do, V. Rosen, and O. Northam. "The Challenges of Phu My 3." Power Engineering International, November, 2002.

"Lumut Combined-cycle Power Plant," published by Segari Energy Ventures Sdn Bhd, Kuala Lumpur, and IMTE, Switzerland, (1997).

Roberts, R., L. Balling, E. Wolt, and M.Fränkle. "The King's Lynn Power Station: The introduction of the advanced single shaft T concept in the IPP market." Power-Gen Europe, Madrid, (June 1997).

"The 347MW King's Lynn Single-Shaft Combined-cycle (GUD) Power Station with Air-Cooled Condenser." Siemens Power Generation publication No. A96001-U10-X-7600.

Index

A

Abu Dhabi Water and Electricity Authority (ADWEA), 371
acid gas removal (AGR)
    alternatives to, 300
    for IGCC, 298–299
    processes for, 297
advanced fast cycling (FACY) capability, 364
Advanced Zero Emission Power plant (AZEP) process, 341–342
aeroderivative gas turbine, 169
    efficiency of, 168
aging, 172
air
    combustion, supply of, 154
    gas turbine historical trends in compressor and flow of, 167
air filter replacement, 174
air pressure. *See* ambient air pressure
air separation unit (ASU). *See also* cryogenic ASU
    future of, 303
    in Puertollano IGCC plant, 383
    Shell gasification and, 310
    use of, 302
air temperature. *See* ambient air temperature
air-cooled blades, 280
air-cooling system. *See* direct air-cooling system
alkali compounds, 295
alpha-value. *See* power coefficient
alternative fuels, 159
alternative working media, 164
ambient air pressure
    operating behavior and, 229
    power output impacted by site elevation and, 54–55

ambient air temperature
    combined-cycle plant efficiency and, 52–53
    evaporative cooling systems and, 246
    gas turbine efficiency and, 51–52
    NO_x emissions and, 262–263
    operating behavior and, 229
    power output impacted by, 53–54
    steam process efficiency impacted by, 52–53
ambient conditions
    gas turbines and, 50
    steam turbines and, 50
ammonia, 3, 271, 296
    bottoming cycle, water mixing with, 164
amortization, 11
    production costs impacted by, 12
Ansaldo Energia, 379
approach temperature, 75
    HRSG and, 84
Arcos III combined-cycle plant
    efficiency of, 367–368
    history of, 365
    multishaft combined-cycle plant configuration of, 366
    technical data of, 367
ash-forming fuels
    contaminants in, 180–181
    ignition problems for, 181–182
automation. *See also* hierarchic levels of automation
    in base load operation, 213
    complexity of, 211
autothermal reforming (ATR), 330
    precombustion CO_2 capture and, 332
availability
    determining, 26–27
    of power plants, 25–26
average electrical efficiency, 14–15

413

B

back-pressure turbine
    cogeneration with, 137–138
    fuel utilization and, 143
balance of plant (BOP), 208
base load operation, 13
    automation in, 213
    electricity costs, power plant comparisons for, 29
biomass power plant. *See* power plants
blast furnace gas, 179
boiler control, 208
bottoming cycle
    ammonia mixing with water in, 164
    defining, 2
    water/steam cycle and, 2–3
BTU natural gas, 175
    characteristics of, 179–180
build-operate-transfer (BOT) project, 357
butane, 179
bypass stacks
    economics and, 208
    gas turbines and, 207
    steam turbines and, 207–208

C

capital cost, 29
    elements impacting, 14
capture and storage (CASTOR), 337
carbon capture and storage (CCS).
    See also Oxyfuel
    demonstration projects for, 346–347
    economics, future estimations for, 345
    efficiency and, 324–325
    EOR and, 325
    studies on, 343–344
carbon dioxide (CO_2) capture. *See also* Oxyfuel; postcombustion CO_2 capture; precombustion CO_2 capture
    efficiency/economics of, 343–344
    IGCC concept based on hard coal, SFG, Siemens combined-cycle and, 314–315
    IGCC concept based on hard coal, Shell gasification, Siemens combined-cycle and, 316–317
carbon dioxide (CO_2) emissions
    combined-cycle plants producing, 273, 390
    from fossil fuels, 323–324
    global warming and, 321
    power plant economics impacted by, 33
    reducing, 323
carbon dioxide (CO_2) separation
    categories of, 325
    from gas mixtures, 325–327
    membranes and, 327
    precombustion CO_2 capture and, 331
    sorption and, 326–327
carbon (CO) shift, 330–331
Carnot efficiency, 53
    calculating, 35
    of combined-cycle plats, 36
    gas turbines and, 36
    of steam turbines, 36
casing. *See* cold-casing design; hot-casing design
CES cycle, 340–341
chemical looping, 328–329
chiller, 245
    advantages of using, 249
    combined-cycle plant power output improvement with, 251–252
    control system for, 252
    economics and, 250
    GT air inlet cooling with, 249–250
    TES with, 251
    typical diagram for, 251
circulation
    HRSG and forced, 184
    HRSG and natural, 185
Claus process, 299, 309
CLINSULF, 299, 313
closed control loops. *See also* drum level control
    in dual-pressure cycle, 219
    feedwater temperature and, 220–221
    live-steam pressure and, 221–222
    live-steam temperature and, 220
    load control/frequency control in, 216–217
    steam turbines, operation of, 222
    types of, 214
closed-circuit air cooling generators, 201–202
closed-steam cooling, 280
clutch, 198–199
coal. *See also* hard coal; shell coal gasification process; steam power plant
    availability of, 9–10
    proved reserves v. yearly consumption of, 10

coal plants, 8. *See also* pulverized coal
  (PC) plants
cogeneration. *See also* Diemen combined-
  cycle cogeneration plant; Monthel
  cogeneration plant
    with back-pressure turbine, 137–138
    categories of, 135
    design for, 144
    dual-pressure cycle HRSG and, 144
    evaluating, 139–143
    with extraction/condensing steam
      turbine, 138
    heat balance for single-pressure cycle with
      supplementary firing and, 140–141
    without steam turbine, 139
    supplementary firing HRSG in, 136
    thermal energy in, 135–136
cold-casing design, 188
combined-cycle plants. *See also* Arcos III
  combined-cycle plant; Diemen combined-
  cycle cogeneration plant; dual-pressure
  cycle; exhaust gas; fully fired combined-cycle
  plant; Monterrey combined-cycle plant;
  Monthel cogeneration plant; multishaft
  combined-cycle plants; Palos de la Frontera
  combined-cycle plant; parallel-fired
  combined-cycle plant; Phu My 3 combined-
  cycle plant; power plants; Puertollano IGCC
  plant; Shuweit S1 Independent Water and
  Power Plant; Siemens combined-cycle;
  single-pressure cycle; single-shaft combined-
  cycle plants; Taranaki combined-cycle plant;
  Vado Liguire power station
    additional equipment for, 209
    advantages of, 3
    alternative fuel applications for, 159
    alternative working media for, 164
    ambient air temperature impacting
      efficiency of, 52–53
    capital requirement breakdown for, 19
    Carnot efficiency of, 36
    chiller improving power output in, 251–252
    CO_2 emissions produced in, 273, 390
    competitive standing of, 17
    concept selection for, 45–46
    control room layout in, 213–214
    conventional steam turbines v., 195–196
    cooling media configurations for, 204,
      207, 389
    defining, 1–2
    degradation typical to, 174–175, 242–243
    design philosophy evolution in, 47
    development trends in, 277–278
    efficiency improvements for, 33–34
    efficiency, with added heat in HRSG of,
      38–40
    efficiency, without added heat in HRSG
      of, 40–41
    emissions from, 262
    environment, impact of, 6–7, 275, 390
    evaporative cooling system for, 204–205
    evaporative cooling system, power output
      increase in, 246–247
    feedwater preheating in, 86
    flow diagram of, 2
    frequency impacting power output in, 232
    fuel flexibility of, 390
    fuel types for, 66
    future trends in, 133–134, 390–391
    gas turbine efficiency and, 234
    gas turbine's application in, 172
    gas turbines converted to, 158
    gas turbines optimized for, 42–44
    global examples of, 349
    installed capacity of, 7
    investors, popularity of, 24
    legislation and selection of, 57–58
    LHV of, 3
    load control droop characteristic in,
      217–218
    off-design behavior corrections for,
      228–234
    off-design behavior of, 225–226
    PFBC applications for, 159–160
    price comparisons for, 18–19
    process steam pressure impacting power
      coefficient/power output in, 141–142
    reheat impacting efficiency of, 133–134
    seawater desalination plants coupled with,
      149–151
    shutdown for, 258–259
    startup time importance to, 196
    startup time ranges of, 253–254
    steam injection, water injection
      impacting, 58–59
    steam process efficiency equation for, 39
    technical standard of, 390
    temperature/entropy diagrams for, 37

testing procedures for, 238–241
thermodynamics of, 35–36, 389
TIT impacting efficiency in, 41–42
waste heat reduced in, 273
water/steam cycle design challenges for, 70
combustion. *See also* oxygen-lean combustion; oxygen-rich combustion; postcombustion CO_2 capture; precombustion CO_2 capture; sequential combustion; staged combustion; stoichiometric combustion
    air supply for, 154
    gas turbines and, 166–167
combustion with excess air, 267–268
Comisión Federal de Electricidad (CFE), 353
compressor
    development of, 165–166
    exit pressure and, 306–307
    fouling in, 172–173
    gas turbine historical trends in air flow of, 167
    improving, 281
    on-line washing/off-line washing for, 173
    power consumed by, 51
    pressure increased by, 65
condenser. *See also* hotwell
    deaeration in, 99–100
    live-steam pressure impacting, 80
    reusing, 153
condenser pressure
    cooling medium temperature v., 62–63
    economics impacted by, 62
condenser vacuum, 56
    efficiency impacted by, 230–231
ConocoPhillips E-Gas gasification process, 294
construction
    economics, interest and time for, 27
    of HRSG without supplementary firing, 183–184
    power plant comparisons on time for, 27–28
contaminants, 180–181
control room, 213–214
control system. *See also* closed control loops; frequency control; load control
    for chiller, 252
    for evaporative cooling system, 252
    features of, 212
    for fogging system, 252
    hierarchic levels of automation in, 212–213

    for hotwell, 222
    necessity of, 208
    for power augmentation, 223
    for process energy, 223
    role of, 211
    for supplementary firing, 223
    for Taranaki combined-cycle plant, 353
    VIGV/TIT and, 215
conventional boiler
    energy/temperature diagram for conventional, 89
    feedwater preheating in, 89
    gas turbine repowering, adapting, 154–155
conversions table, 403
converter gas, 179
cooling media. *See also* closed-steam cooling; GT air inlet cooling; steam cooling
    combined-cycle plant configurations for, 204, 207, 389
    condenser pressure v. temperature of, 62–63
    dry cooling system with ACC for, 61
    options for, 60
    reusing, 153
    temperature impacting, 61–62
    TIT chronology aided by, 279
    water for, 60
cooling water temperature, 59, 204–206
    operating behavior and, 230–231
CoP E-Gas IGCC plant, 318
COREX gas, 179
correction factors, 239
corrosion. *See also* low-temperature corrosion
    alkali compounds causing, 295
    gas turbines impacted by, 175
costs. *See* capital cost; electricity costs; fuel costs; investment costs; operation and maintenance costs; production costs
credit, 11
    power plants and, 49
cryogenic ASU
    IGCC, process of, 302–303
    oxygen generation and, 327–328
customer requirements, 45
    determining, 48–49
customization, 48
cycle concept, 60
cycle performance summary, 130–131

D

damper, 207
deaeration
    in condenser, 99–100
    importance of, 98
    leakages and, 98
    types of, 99
deaerator, 91–92
    eliminating, 99–100
debt-to-equity ratio, 14
    IPPs and, 12
dedusting, 296–297
degradation. *See also* aging; corrosion; fouling
    combined-cycle plant, typical, 174–175, 242–243
    gas turbines, causes of, 172, 242–243
    maintenance for limiting, 285–286
    performance value impacted by, 241–242
    preventing, 173–174
deregulation
    electric power market and, 5–6
    reliability and, 26
desalination plants, 149. *See also* multistage flash desalination units; seawater desalination plants
    Shuweit S1 IWPP as, 373–374
design. *See also* cold-casing design; drum-type design; hot-casing design; off-design behavior; once-through design
    for cogeneration, 144
    combined-cycle plants, evolution of, 47
    combined-cycle plants, water/steam cycle challenges for, 70
    of Diemen combined-cycle cogeneration plant, 368
    district-heating power plants, criteria for, 146
    for economizer, 191–192
    HRSG, horizontal/vertical comparisons for, 185–186
    HRSG, optimizing, 190
    PFBC, complexities of, 160
    of Puertollano IGCC plant, 382
    single-pressure cycle, parameters of, 77–82
    of Taranaki combined-cycle plant, 350
    of Vado Liguire power station, 378–379
design point, 227–228
development time, 28

Diemen combined-cycle cogeneration plant
    design of, 368
    efficiency of, 369–370
    general arrangement of, 371
    process diagram of, 369
    technical data of, 370
dilution, 305–306
direct air-cooling system (ACC)
    arrangement of, 204
    cooling media using dry cooling system with, 61
dirty fuels, 287
district-heating power plants, 135
    design criteria for, 146
    extraction/condensing steam turbines in, 148
    heat balance for dual-pressure cycle and, 147–148
    process of, 145
    steam extraction, pressure stages in, 145–146
drive level, 212
droop characteristic, 217–218
drum level control, 219
drum-type design, 188
    evaporation in., 187
dry cooling system, 204
    cooling media using ACC and, 61
dual-pressure cycle
    closed control loops in, 219
    cogeneration with HRSG in, 144
    district-heating power plant's heat balance for, 147–148
    energy flow diagram for, 97
    energy/temperature diagram for HRSG in, 98
    exhaust temperature, single-pressure cycle v., 69–70
    heat balance for, 96
    for high sulfur fuels, 93–94
    live-steam pressure in, 100–102
    live-steam temperature in, 102–103
    for low sulfur fuels, 95–97
    moisture content, triple-pressure cycle compared to, 113
    pinch point of HP/LP evaporator in, 104–105
    steam turbines, single-pressure cycle v., 96–97

dual-pressure reheat cycle
  energy flow diagram for, 123–124
  performance benefit in, 123
dynamic behavior, 226

E

economics
  bypass stacks and, 208
  CCS, future estimations for, 345
  chiller and, 250
  CO_2 capture and, 343–344
  CO_2 emissions impacting power plants and, 33
  condenser pressure impacting, 62
  construction time and, 27
  environmental restraints impacting, 277
  higher live-steam pressure and, 80–81
  main-steam parameters increased and, 282
  operational flexibility and, 284
  sliding pressure and, 226–227
  STIG cycles and, 161–162
  TIT increases and, 278–279
  of turbo STIG cycle, 162
economizer
  design for, 191–192
  role of, 74
  temperature changes in, 88–89
efficiency. *See also* average electrical efficiency; Carnot efficiency; gas turbine efficiency; net efficiency; steam process efficiency
  of aeroderivative gas turbine, 168
  ambient air temperature impacting combined-cycle plant and, 52–53
  of Arcos III combined-cycle plant, 367–368
  CCS and, 324–325
  closed-steam cooling aiding, 280
  CO_2 capture and, 343–344
  combined-cycle plants, improving, 33–34
  combined-cycle plants, TIT impacting, 41–42
  of combined-cycle plants with added heat in HRSG, 38–40
  of combined-cycle plants without added heat in HRSG, 40–41
  condenser vacuum impacting, 230–231
  cycle performance summary for, 130–131
  of Diemen combined-cycle cogeneration plant, 369–370
  fuel costs, importance of, 19–20
  gas turbines, pressure ratio impacting, 41–42, 65
  of HAT cycle, 163
  of HTW process, 310–311
  of IGCC concepts, 313
  LP impacting HRSG's, 101–102
  of Monterrey combined-cycle plant, 354
  of Monthel cogeneration plant, 385
  of Oxyfuel, 344
  of Palos de la Frontera combined-cycle plant, 361–362
  part-load behavior improvements for, 237
  reheat impacting combined-cycle plants and, 133–134
  repowering increasing, 151
  of Shell gasification, 317
  steam cooling aiding, 280–281
  steam injection impacting, 57–58
  steam turbine equations for, 399
  temperature, after supplementary firing, impacting, 128, 130
  water injection, problems with, 266–267
ELCOGAS, 382
electric power market
  deregulation and, 5–6
  priorities of, 5
  production costs, risks in, 6
electrical corrections, 231–233
electricity
  power plant's type impacted by value of, 13
  production cost of, 14–17
  storage challenges with, 5
  world net generation, by type, of, 8
electricity costs
  calculating, 28–29, 33
  equivalent utilization time impacting, 31–32
  fuel costs impacting, 30–31
  interest rate impacting, 31–32
  power plant comparisons, base load operation and, 29
  power plant comparisons, intermediate load operation and, 30
Electricity of Vietnam (EVN), 357
electricity power demand, 48

Index • 419

emissions, 6. *See also* carbon dioxide (CO_2) emissions; nitrous oxide (NO_x) emissions; SO_x emissions
    combined-cycle plants and, 262
    of IGCC, 289
    legislation limiting, 57
    Switzerland legislation on, 387
    types of, 261
    Vado Liguire power station and, 381
energetic loses
    exergetic loses compared to, 35
    steam turbines, exergetic losses v., 78–79
energy consumption, 327, 331
    global warming, rise in, 321–322
energy/temperature diagram
    for conventional boiler, 89
    for dual-pressure cycle HRSG, 98
    for heat exchanger, 71
    for HRSG and supplementary firing, 126–127
    for single-pressure cycle HRSG, 75, 79, 88
    for triple-pressure cycle HRSG, 108
    for triple-pressure reheat cycle HRSG, 119–120
engineering procurement and construction contract (EPC), 13, 47
Enhanced CO_2 Capture (ENCAP) project, 334
enhanced oil recovery (EOR), 337
    CCS and, 325
entrained-flow gasifers, 290–292
environment. *See also* global warming
    combined-cycle plants' impact on, 6–7, 275, 390
    economics impacted by restraints of, 277
    exhaust gas impacting, 261–262
    IGCC and, 287
    leakages impacting, 164
    waste heat form impacting, 274
equivalent utilization time
    determining, 16
    electricity costs impacted by, 31–32
European Technology Platform for Zero Emission Fossil Fuel Power Plants (ETP-ZEP), 344
evaporation, 149
    once-through design/drum-type design and, 187
evaporative cooling systems
    ambient air temperature and, 246
    for combined-cycle plants, 204–205
    combined-cycle plants, power output increase in, 246–247
    control system for, 252
    fogging system compared to, 247
    goal of, 246
    plume abatement concept for, 206
evaporator
    dual-pressure cycle, pinch point in HP/LP and, 104–105
    role of, 74
    triple-pressure cycle, pinch point in HP/LP and, 114
evaporator loop, 93
    feedwater preheating with, 91–92
exergetic loses
    energetic loses compared to, 35
    steam turbines, energetic loses v., 78–79
exergy transfer
    improving, 123
    in triple-pressure reheat cycle, 117
exhaust gas, 74–75
    environment impacted by, 261–262
    evaluating, 131–132
    HRSG, pressure loss in, 84–85
    water/steam cycle's heat transfer v., 91
exhaust losses, 400
exhaust temperature
    dual-pressure cycle v. single-pressure cycle and, 69–70
    gas turbines impacted by, 68
exit pressure, 306–307
extraction/condensing steam turbines
    cogeneration with, 138
    in district-heating power plants, 148

F

far field noise, 275
fast cycling provisions, 285. *See also* advanced fast cycling (FACY) capability
feedwater preheating
    combined-cycle plant and, 86
    conventional boiler and, 89
    evaporator loop for, 91–92
    HRSG impacted by, 86–88
    recirculation and, 90
    in steam power plant, 86
    steam turbines impacted by, 94

feedwater temperature
    closed control loops and, 220–221
    HRSG and control for, 221
financing structure
    creating, 14
    terms of, 13
finned tubing, 189–190
fixed costs, 17
    power plants, O&M costs and, 25
fixed-bed processes, 290–291
flame temperature
    NO_x emissions, gas turbines and, 263–264
    in stoichiometric combustion, 264
flashing, 315
    in seawater desalination plants, 149
flexibility. *See* operational flexibility
FLEXSORB unit, 313
fluidized-bed (FB) processes, 290–291
fly ash, 296–297
fogging system, 245. *See also* high-fogging system
    control system for, 252
    evaporative cooling system compared to, 247
    principles of, 247
    typical arrangement, 248
fossil fuels
    CO_2 emissions from, 323–324
    consumption of, 8
    price volatility of, 23
    proved reserves v. yearly consumption for, 10
fouling
    in compressor, 172–173
    of turbines, 174
frequency
    combined-cycle power output impacted by, 232
    operating behavior and, 231–232
frequency control, 216
    closed control loops and, 216–217
    load control compared to, 217
front-end engineering design (FEED) studies, 319
fuel. *See also* alternative fuels; dirty fuels; gaseous fuels; high sulfur fuels; liquid fuels; low sulfur fuels
    availability of, 9–10
    combined-cycle plant, flexibility of, 390
    combined-cycle plant, types of, 66
    critical properties of, 176
    for gas turbines, 175–176
    LHV's importance in, 64, 233–234
    off-design behavior and, 233
    power plants flexibility of, 22–23
    power plants, selecting, 21
    preheating, 64–65
    proved reserves v. yearly consumption of, 10
fuel changeover
    diagram for, 260
    duration of, 259
    purpose of, 258
    steps for, 259–260
fuel costs
    average electrical efficiency and, 14–15
    efficiency's importance to, 19–20
    electricity costs impacted by, 30–31
    volatility of, 23–24
fuel utilization
    back-pressure turbine and, 143
    determining, 139
    single-pressure cycle, power coefficient impacting, 142–143
fuel-to-air ratio, 263–265
fully fired combined-cycle plant, 154–155
functional group level, 212

G

gas mixtures, 325–327
gas turbine efficiency
    ambient air temperature and, 51–52
    combined-cycle plants and, 234
    increasing, 40
    part-load behavior and, 235
    steam process efficiency impacting, 41
gas turbines. *See also* aeroderivative gas turbine; GT air inlet cooling; heavy-duty industrial gas turbines; Siemens SGT5-8000H gas turbine; steam-injected gas turbine (STIG) cycles; turbine inlet temperature
    ambient conditions and, 50
    bypass stacks and, 207
    capacity improvements of, 284
    Carnot efficiency of, 36
    categories of, 168
    combined-cycle plants, application of, 172
    combined-cycle plants converted from, 158

combined-cycle plants with optimized, 42–44
combustion and, 166–167
conventional boiler adaptation for repowering, 154–155
corrosion impacting, 175
degradation causes in, 172, 242–243
exhaust temperature impacting, 68
fuel for, 175–176
high-fogging system impacting, 248–249
HRSG's function for, 183
ignition and, 254–255
inspections for, 172
load jumps and, 218
multi-shaft v. single shaft, 67
NO_x emissions, flame temperature in, 263–264
plant concept solution, selecting, 66–70
power output data for, 171
pressure ratio impacting efficiency in, 41–42, 65
price comparisons for, 18–19
process of, 165
reheat impacting, 114
sequential combustion and, 170–171
sequential combustion's advantages for, 44
in single-pressure cycle, 171
startup quickness of, 253
syngas utilization in modified, 305–306
TIT, compressor air flow historical trends in, 167
TIT increased in development of, 165–166
turbine mass flow imbalance in, 306–307
gaseous fuels
typical composition of, 180
Wobbe index for comparing, 177
gasification. *See also* high temperature winkler (HTW) process; Integrated gasification combined-cycle plants; partial oxidation (POX) gasification
basic technologies for, 290–292
ConocoPhillips E-Gas gasification process for, 294
GE coal gasification process for, 293
history of, 290
MHI dry-feed gasifer for, 294
SCGP for, 292–293
SFG and, 293
syngas process in, 288

GE coal gasification process, 293
GE energy, IGCC concept, 318
geared high-pressure turbine, 282–283
generator losses, 400
generators. *See* heat recovery steam generator
GE's SPEEDTRONIC, 367
global warming, 1
CO_2 emissions and, 321
energy consumption impacting, 321–322
natural gas mitigating, 324
NO_x emissions and, 262
Graz cycle, 339–340
GT air inlet cooling
chiller impacting, 249–250
investment costs of, 253
overview of, 246
types of, 245

H

hard coal
IGCC concept based on CO_2 capture, SFG, Siemens combined-cycle and, 314–315
IGCC concept based on CO_2 capture, Shell gasification, Siemens combined-cycle and, 316–317
heat balance
for cogeneration, single-pressure cycle with supplementary firing, 140–141
for district-heating power plant with dual-pressure cycle, 147–148
for dual-pressure cycle, 96
performance guarantees, computer model for, 240
for single-pressure cycle, 76
for single-pressure cycle with supplementary firing, 129
for triple-pressure cycle, 106–107
for triple-pressure reheat cycle, 116–117
heat exchanger(s). *See also* economizer; evaporator; superheater
energy exchange in idealized, 70–71
energy/temperature diagram for, 71
HRSG, equations for, 393–395
HRSG, sections of, 74
heat recovery steam generator (HRSG). *See also* economizer; evaporator; superheater
approach temperature and, 84

cogeneration with dual-pressure cycle
and, 144
cogeneration with supplementary firing
in, 136
cold-casing design for, 188
combined-cycle plants' efficiency, added
heat in, 38–40
combined-cycle plants' efficiency, without
added heat in, 40–41
construction without supplementary
firing for, 183–184
contradictory conditions for, 189
design, horizontal/vertical comparisons
for, 185–186
design optimization for, 190
dual-pressure cycle, energy/temperature
diagram for, 98
energy/temperature diagram for
supplementary firing and, 126–127
exhaust gas pressure loss in, 84–85
feedwater preheating impacting, 86–88
feedwater temperature control and, 221
finned tubing for, 189–190
forced circulation in, 184
gas turbines, function of, 183
heat exchanger equations for, 393–395
heat exchanger sections of, 74
heat transfer coefficient for, 395–398
heat transfer factors for, 71–72
heat utilization improved in, 64–65
horizontal type of, 185
hot-casing design for, 188–189
limited supplementary firing in, 192–194
low-temperature corrosion in, 90–91, 190
LP impacting efficiency of, 101–102
maximum supplementary firing in, 194–195
natural circulation in, 185
once-through design for, 186–188
in Palos de la Frontera combined-cycle
plant, 363
pinch point determining heating surface
of, 75, 83–84
pinch points in modern, 190
with SCR, 271–272
in single-pressure cycle, 74–75
single-pressure cycle, energy/temperature
diagram for, 75, 79, 88
startup challenges with, 190–191

supplementary firing and, 125
temperature control and, 220
triple-pressure cycle, energy/temperature
diagram for, 108
triple-pressure reheat cycle, energy/
temperature diagram for, 119–120
vertical type of, 184
heat transfer coefficient, 395–398
heavy-duty industrial gas turbines, 170
developments in, 169
power output of, 168–169
Henry's law, 326
hierarchic levels of automation, 212–213
high sulfur fuels, 91–92
dual-pressure cycle for, 93–94
high temperature winkler (HTW) process
efficiency of, 310–311
IGCC concept based on lignite, Siemens
combined-cycle and, 311–313
high volatile liquid fuels, 182
higher heating value (HHV), 313
determining, 20
high-fogging system
gas turbines impacted by, 248–249
potential of, 252–253
power augmentation increased in, 248
high-pressure (HP), 57
dual-pressure cycle, pinch point in
evaporator with, 104–105
pinch point, LP's relation to, 105
steam turbine power output impacted by
LP and, 100–101
triple-pressure cycle, pinch point in
evaporator with, 114
hot-casing design, 188–189
hotwell
control system for, 222
steam turbines and, 99
HT sour shift, 300
humid air turbine (HAT) cycle
challenges of, 164
efficiency of, 163
humidity. *See* relative humidity
hybrid cooling tower, 206
hydrogen, 177–179, 305
precombustion CO_2 capture and behavior
of, 333
hydrogen-cooled generators, 201

I

ignition
  ash-forming fuels, problems with, 181–182
  in gas turbines, 254–255
independent power producers (IPPs)
  debt-to-equity ratio and, 12
  risks and, 11
  SPC used by, 13
indices, 406
indirect air-cooling system, 204
industrial power stations, 135
  steam extraction in, 136–137
inlet guide vane (IGV), 258
inspections
  gas turbines and, 172
  intervals between, 168
installed capacity
  of combined-cycle plants, 7
  power plant evolution in, 9
  of steam turbines, 7
integrated CO_2 capture. *See* Oxyfuel
integrated gasification combined-cycle plants (IGCC), 1. *See also* CoP E-Gas IGCC plant; Puertollano IGCC plant; Zero Emission Integrated Gasification Combined Cycle
  AGR processes for, 298–299
  CO_2 capture, hard coal, SFG, Siemens combined cycle concept for, 314–315
  CO_2 capture, hard coal, Shell gasification, Siemens combined cycle concept for, 316–317
  concepts for, 308
  cryogenic ASU process in, 302–303
  demonstration plants for, 289–290
  efficiency/investment cost of concepts for, 313
  emissions of, 289
  environment and, 287
  GE energy concept for, 318
  investment costs of, 289, 319–320
  lignite, HTW process, Siemens combined cycle concept for, 311–313
  Rectisol/Selexol and, 302
  Shell gasification, Siemens combined-cycle concept for, 308–310
  syngas conditioning system in, 304

Integrated Reforming Combined Cycle (IRCC), 331–332
interest
  construction time and, 27
  production costs impacted by, 12
interest rate, 13–14
  electricity costs impacted by, 31–32
intergovernmental Panel of Climate Change (IPCC), 321
intermediate load operation, 13
  electricity costs, power plant comparisons for, 30
intermediate-pressure (IP), 57
internal rate of return (IRR), 11
investment costs
  elements impacting, 14
  of GT air inlet cooling, 253
  of IGCC, 289
  of IGCC concepts, 313, 319–320
  PFBC, complexities with, 160
  of Shell gasification, 317
investors
  combined-cycle plants' popularity with, 24
  targets of, 11
ion transport membranes, 328
Italy. *See* Vado Liguire power station

J–K

Joule and Clausius-Ranking cycle (combined cycle), 323–324

L

Law of Cones, 400–402
leakages
  deaeration and, 98
  environment impacted by, 164
legislation, 261
  combined-cycle plant selection and, 57–58
  emissions limited in, 57
  NO_x emissions regulated by, 270–271
  of Switzerland on emissions, 387
light distillate fuel, 175
  characteristics of, 180
lignite, 311–313

liquefied natural gas (LNG), 19
    characteristics of, 178
liquefied petroleum gas (LPG), 177, 179
liquid fuels, 180–183
live-steam pressure
    closed control loops and, 221–222
    condenser impacted by, 80
    dual-pressure cycle and, 100–102
    economic advantages of higher, 80–81
    moisture content and, 79
    part-load behavior and, 236–237
    steam turbines, effect of, 77–78, 199–200
    triple-pressure cycle and, 110–111
    triple-pressure reheat cycle and, 120–121
live-steam temperature
    closed control loop and, 220
    dual-pressure cycle and, 102–103
    part-load behavior and, 236–237
    steam turbines, impact of increasing, 81–82
    triple-pressure cycle and, 111–112
    triple-pressure reheat cycle and, 122
load control
    closed control loops and, 216–217
    combined-cycle plant, droop characteristic of, 217–218
    frequency control compared to, 217
    supplementary firing and, 217
load factor
    measuring, 16
    purpose of, 14
load jump, 225
    gas turbines and, 218
LO-CAT, 300
long-term sales contracts (LTSCs), 11
low boiling liquid fuels, 182
low sulfur fuels, 95–97
lower heating value (LHV)
    of combined-cycle plants, 3
    determining, 20
    fuel, importance of, 64, 233–234
low-NO_x burner. *See also* Siemens dry low-NO_x burner
    characteristics of, 268
    cross section of, 269
low-pressure (LP), 57
    dual-pressure cycle, pinch point in evaporator with, 104–105
    HRSG efficiency impacted by, 101–102
    pinch point, HP's relation to, 105
    steam turbine power output impacted by HP and, 100–101
    triple-pressure cycle, pinch point in evaporator with, 114
low-temperature corrosion, 90–91, 190
LT sweet shift, 300

M

machine level, 212
main-steam parameters
    economics of increasing, 282
    geared high-pressure turbine and, 282–283
    historical improvement of, 281–282
maintenance, 285–286. *See also* operation and maintenance costs
makeup water, 100
market price of electricity, 11
mechanical draft cooling tower, 205
mechanical losses, 400
membranes, 327. *See also* ion transport membranes
methanol, 182–183
Mexico. *See* Monterrey combined-cycle plant
MHI dry-feed gasifer, 294
moisture content, 78
    live-steam pressure impacting, 79
    reheat and, 115
    triple-pressure cycle compared to dual-pressure cycle for, 113
    triple-pressure reheat cycle decreasing, 118
Monterrey combined-cycle plant
    CFE ordering of, 353
    composition of, 354
    efficiency of, 354
    layout of, 356–357
    performance data for, 355–356
    process diagram for, 355
Monthel cogeneration plant
    arrangement of, 386–387
    efficiency of, 385
    process diagram for, 386
    technical data of, 387
multishaft combined-cycle plants
    Arcos III combined-cycle plant configuration as, 366

configuration for, 197
Phu My 3 combined-cycle plant as, 358–359
multistage flash desalination units (MSF), 373

N

natural draft cooling tower, 205
natural gas, 175. *See also* liquefied natural gas
    availability of, 9–10
    global warming mitigation with, 324
    pressure fluctuations with, 177–178
    proved reserves v. yearly consumption of, 10
natural gas premix burner, 333–334
near field noise, 275
net efficiency
    defining, 20
    power plant comparisons on, 20–21
net present value (NPV), 282
Netherlands. *See* Diemen combined-cycle cogeneration plant
New Zealand. *See* Taranaki combined-cycle plant
nitrous oxide (NO_x) emissions. *See also* low-NO_x burner
    ambient air temperature and, 262–263
    combustion with excess air reducing, 267–268
    fuel-to-air ratio and, 263–265
    gas turbines, flame temperature and, 263–264
    global warming and, 262
    legislation regulating, 270–271
    reducing, 57
    SCR disadvantages for reducing, 271
    staged combustion reducing, 267
    syngas dilution controlling, 305–306
    water injection/steam injection and, 265–266
noise emissions, 261
    types of, 275
nuclear power plant. *See* power plants
nuclear waste, 261

O

off-design behavior
    combined-cycle plants and, 225–226
    combined-cycle plants, corrections for, 228–234
    fuel and, 233
off-line washing, 173
oil
    availability of, 9–10
    proved reserves v. yearly consumption of, 10
oil-soluble contaminants, 181
once-through design, 188
    evaporation in, 187
    HRSG with, 186–188
once-through water cooling, 204
    configuration/role of, 206
on-line washing, 173
open cycle gas turbine, 172
    advantages of, 3
open-circuit air cooling generators, 201
operating behavior
    ambient air pressure and, 229
    ambient air temperature and, 229
    calculating, 226–227
    cooling water temperature and, 230–231
    design point solution for calculating, 227–228
    electrical corrections for, 231–233
    frequency impacting, 231–232
    parameters for correcting, 228
    power factor and, 232
    process energy and, 233
    relative humidity and, 229–230
operation and maintenance (O&M) costs, 15, 29
    power plants, fixed costs and, 25
    power plants, variable costs and, 24
operational flexibility
    economics and, 284
    fast cycling provisions for, 285
original equipment manufacturers (OEMs), 285–286
overpressure deaeration, 99
Oxyfuel, 327–328
    AZEP process for, 341–342
    CES cycle concept for, 340–341
    direct/indirect systems in, 339
    efficiency of, 344

Graz cycle concept for, 339–340
processes of, 338
ZESOFC concept for, 342–343
oxygen generation
chemical looping and, 328–329
cryogenic ASU and, 327–328
ion transport membranes and, 328
oxygen-lean combustion, 267
oxygen-rich combustion, 267

P–Q

Palos de la Frontera combined-cycle plant
efficiency of, 361–362
HRSGs in, 363
layout of, 362
location of, 361
technical data of, 364
parallel-fired combined-cycle plant
operating modes for, 157–158
reheat in, 156
repowering and, 155–156
steam turbine converted to, 157
partial oxidation (POX) gasification, 330
part-load behavior
calculating, 226–227
efficiency improvements for, 237
gas turbine efficiency with, 235
live-steam pressure/live-steam
temperature with, 236–237
reheat and, 236
single-shaft combined-cycle plants and, 237–238
peak load operation, 13
performance guarantees
comparison for, 241
heat balance computer model for, 240
measuring, 238–239
power output correction factors for, 239
performance monitoring, 244
performance test, 241
performance value
degradation impacting, 241–242
monitoring, 244

petcoke, 289
Phu My 3 combined-cycle plant
as BOT project, 357
location of, 358
as multishaft combined-cycle plant, 358–359
Siemens combined-cycle and
arrangement of, 359
technical data of, 360
pinch point
in HP/LP evaporator, dual-pressure cycle, 104–105
in HP/LP evaporator, triple-pressure cycle, 114
HP/LP interrelation in, 105
in HRSG, 190
HRSG heating surface determined by, 75, 83–84
steam turbine power output, impact of, 83–84
piston losses, 399
plant concept solution, 45, 132
gas turbine selection for, 66–70
plume abatement, 206
population growth, 321
postcombustion CO_2 capture
methods for, 335
SARGAS cycle for, 337–338
sorption and, 335–336
sorption systems for, 337
power augmentation
applications for, 245
control system for, 223
goal of, 244
high-fogging system increasing, 248
power coefficient
determining, 139–140
parameters impacting, 141
process steam pressure impacting
combined-cycle plants and, 141–142
single-pressure cycle, fuel utilization
impacted by, 142–143
supplementary firing impacting, 143
power factor, 232
power island, 67
benefits of, 47–48

power output
  ambient air pressure, site elevation impacting, 54–55
  ambient air temperature impacting, 53–54
  chiller improving combined-cycle plant and, 251–252
  cycle performance summary for, 130–131
  evaporative cooling system in combined-cycle plant increasing, 246–247
  frequency impacting combined-cycle plants and, 232
  gas turbines, data for, 171
  of heavy-duty industrial gas turbines, 168–169
  HP, LP impacting steam turbines and, 100–101
  performance guarantees, correction factors for, 239
  pinch point impacting steam turbines and, 83–84
  process steam pressure impacting combined-cycle plants and, 141–142
  relative humidity and, 56
  steam injection, water injection impacting, 58
  of steam turbines, 72, 399
  temperature, after supplementary firing, impacting, 128, 130
power plants. *See also* combined-cycle plants; district-heating power plants; gas turbines; industrial power stations; seawater desalination plants; steam power plant; steam turbines
  availability of, 25–26
  base load operation, electricity cost comparisons for, 29
  CO_2 emissions impacting economics of, 33
  construction time comparisons for, 27–28
  credit and, 49
  electricity value impacting choice of, 13
  fuel flexibility of, 22–23
  fuel selection considerations for, 21
  installed capacity evolution of, 9
  intermediate load operation, electricity cost comparisons for, 30
  net efficiency comparisons for, 20–21
  O&M costs, fixed costs for, 25
  O&M costs, variable costs for, 24
  present value determined for, 15–16
  reliability of, 25–26
  risks for, 12
  as trading tool, 225
  waste heat dissipation in various, 274
precombustion CO_2 capture, 329
  ATR and, 332
  CO_2 separation and, 331
  hydrogen behavior in, 333
  syngases and, 332
preheating. *See also* feedwater preheating
  fuel and, 64–65
  increasing stages of, 94
present value, 15–16
pressure. *See also* ambient air pressure; condenser pressure; dual-pressure cycle; high-pressure; intermediate-pressure; live-steam pressure; low-pressure; process steam pressure; single-pressure cycle; triple-pressure cycle
  compressor increasing, 65
  HRSG exhaust gas, loss in, 84–85
  natural gas, fluctuations in, 177–178
  steam extraction, district-heating power plants, stages of, 145–146
  superheater, water/steam cycle drop in, 84
pressure ratio, 51, 165
  gas turbine efficiency impacted by, 41–42, 65
pressurized fluidized-bed combustion (PFBC)
  combined-cycle plant applications with, 159–160
  design/investment cost complexities for, 160
process energy
  control system for, 223
  operating behavior and, 233
process steam pressure, 141–142
production costs
  amortization, interest impacting, 12
  electric power market and, 6
  of electricity, 14–17
propane, 179
Puertollano IGCC plant
  ASU in, 383
  availability issues of, 384
  design of, 382
  technical data of, 384
pulverized coal (PC) plants, 288
purge time, 254–255

R

radioactivity, 261
recirculation
    feedwater preheating with, 90
    loop for, 124
Rectisol, 302
reheat. *See also* dual-pressure reheat cycle; triple-pressure reheat cycle
    combined-cycle efficiency impacted by, 133–134
    gas turbines impacted by, 114
    moisture content and, 115
    in parallel-fired combined cycle plant, 156
    part-load behavior and, 236
    steam turbine expansion line impacted by, 113, 118
relative humidity
    operating behavior and, 229–230
    power output and, 56
reliability
    deregulation and, 26
    determining, 26–27
    of power plants, 25–26
repowering
    efficiency increased in, 151
    with fully fired combined-cycle plant, 154–155
    gas turbines, conventional boiler adaptation for, 154–155
    in parallel-fired combined-cycle plant, 155–156
    steam turbine before/after, 151–152
    steam turbine, problems with, 153–154
    steam turbines, equipment reuse in, 153
    Vado Liguire power station, thermal process after, 379–380
resources, 60. *See also* cooling media; fuel
return on equity (ROE), 11
risks
    electric power market and, 6
    IPPs mitigating, 11
    power plants and, 12

S

SARGAS cycle, 337–338
seawater desalination plants, 135
    combined-cycle plants coupled with, 149–151
    flashing in, 149
selective catalytic reduction (SCR), 271–272
SELEXOL, 298
    IGCC and, 302
separator, 123
sequential combustion
    gas turbine's advantages with, 44
    gas turbines with, 170–171
shell coal gasification process (SCGP), 292–293
Shell gasification
    ASU for, 310
    diagram of, 311
    efficiency/investment costs of, 317
    IGCC concept based on CO_2 capture, hard coal, Siemens combined-cycle and, 316–317
    IGCC concept based on Siemens combined-cycle and, 308–310
shutdown, 258–259
Shuweit S1 Independent Water and Power Plant (IWPP)
    as desalination plant, 373–374
    functioning of, 372–373
    location of, 371–372
    technical data of, 374
Siemens combined-cycle
    IGCC concept based on CO_2 capture, hard coal, SFG and, 314–315
    IGCC concept based on CO_2 capture, hard coal, Shell gasification and, 316–317
    IGCC concept based on lignite, HTW process and, 311–313
    IGCC concept based on Shell gasification and, 308–310
    Phu My 3 combined-cycle plant arrangement and, 359
Siemens dry low-NO_x burner, 270
Siemens fuel gasifer (SFG)
    gasification with, 293
    IGCC concept based on CO_2 capture, hard coal, Siemens combined-cycle and, 314–315
Siemens SGT5-8000H gas turbine, 284

single-line diagram, 202–203
single-pressure cycle
　design parameters of, 77–82
　energy flow diagram for, 77
　energy/temperature diagram for HRSG in, 75, 79, 88
　example of, 76–77
　exhaust temperature, dual-pressure cycle v., 69–70
　flow diagram for, 73–74
　gas turbines in, 171
　heat balance for, 76
　heat balance for cogeneration with supplementary firing in, 140–141
　heat balance for supplementary firing in, 129
　HRSG in, 74–75
　power coefficient impacting fuel utilization in, 142–143
　steam turbines, dual-pressure cycle v., 96–97
single-shaft combined-cycle plants
　clutch and, 198–199
　configuration for, 197–198
　cross section of, 199
　part-load behavior and, 237–238
　single-line diagram for, 202–203
　steam turbines in, 198
SINTEF, 344
site conditions, 45
site elevation, 54–55
site-related factors, 45. *See also* ambient air pressure; ambient air temperature; ambient conditions; relative humidity
　location and, 49
sliding pressure, 226–227
solid oxide fuel cell (SOFC), 342
sorption
　CO_2 separation and, 326–327
　postcombustion CO_2 capture and, 335–336
　postcombustion CO_2 capture, systems for, 337
SO_x emissions, 272
Spain. *See* Arcos III combined-cycle plant; Palos de la Frontera combined-cycle plant; Puertollano IGCC plant
Special purpose company (SPC), 13
stack loses, 109
staged combustion, 267

standardization, 132, 361
　customization compared to, 48
standstill
　startup, 8 hours, 256
　startup, 48 hours, 257
　startup, 120 hours, 257
　startup and, 254
startup. *See also* ignition
　8 hours standstill, curve for, 256
　48 hours standstill, curve for, 257
　120 hours standstill, curve for, 257
　combined-cycle plants and, 196
　combined-cycle plants' range of times for, 253–254
　gas turbines, quickness of, 253
　HRSG, challenges with, 190–191
　procedure for, 254–256
　standstill impacting, 254
　for steam turbines, 255–256
　times/values for, 255
steam bypass, 221–222
steam cooling, 280–281
steam extraction
　cycle with, 138
　district-heating power plants, pressure stages for, 145–146
　industrial power stations and, 136–137
steam injection
　combined-cycle plant, impact of, 58–59
　efficiency and, 57–58
　NO_x emissions and, 265–266
　power output impacted by, 58
steam parameters. *See* main-steam parameters
steam power plant. *See also* power plants
　feedwater preheating in, 86
　price comparisons for, 18–19
steam process efficiency
　ambient air temperature impacting, 52–53
　combined-cycle plant, equation for, 39
　gas turbine efficiency impacting, 41
steam reforming, 330
steam turbine expansion line
　reheat impacting, 113, 118
　in triple-pressure reheat cycle, 122
steam turbines. *See also* back-pressure turbine; extraction/condensing steam turbines; water/steam cycle
　ambient conditions and, 50
　bypass stack and, 207–208

Carnot efficiency of, 36
characteristics of, 196
closed control loops operation for, 222
cogeneration with no, 139
combined-cycle plants v. conventional, 195–196
efficiency equations for, 399
energetic loses v. exergetic loses in, 78–79
exhaust losses for, 400
feedwater preheating impacting, 94
generator losses in, 400
hotwell in, 99
HP, LP impacting power output of, 100–101
installed capacity of, 7
live-steam pressure impacting, 77–78, 199–200
live-steam temperature impacting, 81–82
mechanical losses for, 400
parallel-fired combined-cycle plant converted from, 157
pinch point impacting power output of, 83–84
piston losses and, 399
power output of, 72, 399
repowering, before/after for, 151–152
repowering, equipment reuse in, 153
repowering problems in, 153–154
single-pressure cycle v. dual-pressure cycle in, 96–97
in single-shaft combined-cycle plants, 198
startup for, 255–256
swallowing capacity of, 398
steam-injected gas turbine (STIG) cycles. *See also* turbo STIG cycle
    economics of, 161–162
    energy flow diagram for, 161
stoichiometric combustion, 166
    flame temperature in, 264
stress controller. *See* turbine stress controller
Sulferox, 300
sulfur, 64, 159. *See also* high sulfur fuels; low sulfur fuels; SO_x emissions
    recovery, 299

superheater
    role of, 74
    simplicity v., 103
    water/steam cycle pressure drop in, 84
supplementary firing
    advantages of, 125
    cogeneration, HRSG with, 136
    control system for, 223
    energy/temperature diagram for HRSG and, 126–127
    heat balance for cogeneration, single-pressure cycle with, 140–141
    heat balance for single-pressure cycle with, 129
    HRSG and, 125
    HRSG construction without, 183–184
    HRSG with limited, 192–194
    HRSG with maximum, 194–195
    load control and, 217
    power coefficient impacted by, 143
    temperature impacting power output, efficiency after, 128, 130
swallowing capacity, 398
sweep gas, 341
Switzerland, 387. *See also* Monthel cogeneration plant
symbols, 405–406
syngas conditioning system, 304
syngases, 176
    advanced cleaning technology for, 300–302
    characteristics of, 179
    cleaning techniques for, 295
    dedusting and, 296–297
    gas turbine modification for utilizing, 305–306
    gasification process with, 288
    NO_x emissions, dilution of, 305–306
    precombustion CO_2 capture and, 332
    water gas shift reaction and, 301
system of equations, 400–402

Index • 431

T

Taranaki combined-cycle plant
    arrangement of, 352–353
    control system for, 353
    design of, 350
    process diagram for, 351
    technical data of, 352
temperature. *See also* ambient air temperature; approach temperature; cooling water temperature; energy/temperature diagram; exhaust temperature; feedwater temperature; flame temperature; high temperature winkler (HTW) process; live-steam temperature; low-temperature corrosion; pinch point
    combined-cycle plants, diagrams on entropy and, 37
    condenser pressure v. cooling media and, 62–63
    cooling media impacted by, 61–62
    economizer, changes in, 88–89
    HRSG and controlling, 220
    power output/efficiency, after supplementary firing impacted by, 128, 130
terawatt hour (TWh), 8
thermal barrier coating (TBC), 279
thermal cycles, 1–2
    consumption breakdown of, 8
    popularity of, 8
thermal energy, 135–136
thermal energy storage (TES), 251
thermal process, 379–380
thermodynamics, 35–36, 389
Tolling Agreement, 12
top gas, 179
topping cycle, 2
trading tool, 225
triple-pressure cycle
    adding, 105
    energy flow diagram for, 105–106, 109
    energy/temperature diagram for HRSG in, 108
    heat balance for, 106–107
    live-steam pressure in, 110–111
    live-steam temperature in, 111–112
    moisture content, dual-pressure cycle compared to, 113
    pinch point of HP/LP evaporator in, 114

triple-pressure reheat cycle
    energy flow diagram for, 116, 119
    energy/temperature diagram for HRSG in, 119–120
    exergy transfer in, 117
    heat balance for, 116–117
    live-steam pressure in, 120–121
    live-steam temperature in, 122
    moisture content decreased in, 118
    performance benefit in, 123
    steam turbine expansion line in, 122
tubing. *See* finned tubing
turbine inlet temperature (TIT), 67–68. *See also* GT air inlet cooling
    chronology of cooling media improving, 279
    combined-cycle plant efficiency impacted by, 41–42
    control system and, 215
    definitions, 38
    economics of increasing, 278–279
    gas turbine development increasing, 165–166
    gas turbine historical trends in, 167
    importance of, 44
turbine mass flow, 306–307
turbine stress controller (TSC), 254
turbo STIG cycle, 162

U

unconverted carbon (char), 296–297
Union Fenosa Generacion (UFG), 361
unit level, 212
United Arab Emirates. *See* Shuweit S1 Independent Water and Power Plant

V

vacuum deaeration, 99
Vado Liguire power station
    design of, 378–379
    emissions in new, 381
    main equipment supply for, 379
    new layout for, 376–377
    pre-existing situation of, 375–376
    repowering, thermal process of, 379–380
    technical data of, 378
vaporized liquids, 179–180
variable costs, 17
    power plants, O&M costs and, 24
variable inlet guide vane (VIGV), 215
Vietnam. *See* Phu My 3 combined-cycle plant

W–Y

waste heat, 261
    combined-cycle plants reducing, 273
    environment impacted by form of, 274
    power plants' dissipation of, 274
waste water, 261
water. *See also* cooling water temperature; makeup water; waste water
    bottoming cycle, ammonia mixing with, 164
    for cooling media, 60

water gas shift reaction, 301
water injection, 57
    combined-cycle plant, impact of, 58–59
    efficiency problems with, 266–267
    NO_x emissions and, 265–266
water-cooled generators, 201
water-soluble contaminants, 181
water/steam cycle
    bottoming cycle and, 2–3
    combined-cycle plant design challenges with, 70
    concepts for, 72–73
    evaluating, 131–132
    exhaust gas' heat transfer v., 91
    superheater, pressure drop in, 84
wet cell cooling tower, 63
Wobbe index, 179
    gaseous fuel comparison with, 177

Z

Zero Emission Integrated Gasification Combined Cycle (ZEIGCC), 331–332
zero emission solid oxide fuel cell (ZESOFC) concept, 342–343

About the Authors

Rolf Kehlhofer is an internationally recognized expert in the energy industry and, more specifically, in the field of combined-cycle power plants. The founder and managing partner of The Energy Consulting Group Ltd. in Zurich, Kehlhofer previously served as president of ABB Power Generation Switzerland. A Swiss citizen, Kehlhofer is a graduate of the Swiss Federal Institute of Technology. During his career, his contributions have resulted in nearly 60,000 megawatts of generated power worldwide.

Bert Rukes joined Siemens in 1978 and has held various leadership positions during his tenure there. He currently oversees plant design and development, and nuclear and steam operations. During his distinguished career, Rukes received the Siemens Top Innovator award for outstanding innovations in fossil power plant solutions. He is a member of VDI (the Association of German Engineers) and is an Honorary Fellow of the International Association of the Properties of Water and Steam. A mechanical engineering graduate at the University of Applied Sciences in Düsseldorf and the Technical University in Berlin, Rukes also holds a doctorate in mechanical engineering.

Frank Hannemann has a degree in Thermal Engineering from the Technical University of Zittau. After ten years with Siemens Power Generation, he became Director of IGCC and innovative power plant cycles with special focus on CO_2 free technologies. He is currently head of engineering at Siemens Fuel Gasification Technology GmbH and is responsible for technology development and engineering activities for gasification projects. Hannemann has written articles for numerous publications about integrated gasification combined-cycle plants (IGCC). He is a holder of several patents and his inventions have been named by Siemens AG as one of the most attractive solutions for future power generation.

Franz J. Stirnimann has held the position of senior consultant at The Energy Consulting Group Ltd. in Switzerland since 2005. He started his career as a process engineer and spent more than 25 years in the field of engineering for thermal power plants, mainly for combined-cycle plants within ABB Power Generation and later ALSTOM Power. In 1994 he became vice president for Combined Cycle Plant Engineering with worldwide responsibilities. During his career, he has been involved in the design and execution of all types of combined-cycle plants as well as development activities for plant standardization. A Swiss citizen, Stirnimann has a degree in mechanical engineering from the Swiss University of Applied Science and Arts in Lucerne.